Edexcel AS Physics Revision Guide

for SHAP and concept-led approaches

REVISION GUIDE

WITHDRAWN

A PEARSON COMPANY

Published by Pearson Education Limited, a company incorporated in England and Wales, having its registered office at Edinburgh Gate, Harlow, Essex, CM20 2JE. Registered company number: 872828

Edexcel is a registered trade mark of Edexcel Limited

Text © Pearson Education Ltd 2009

The rights of Pauline Anning, Keith Bridgeman, Richard Laird, Penny Johnson and Tim Tuggey have been asserted by them in accordance with the Copyright, Designs and Patents Act of 1988.

First published 2009

12 11 10 09
10 9 8 7 6 5 4 3 2 1

British Library Cataloguing in Publication Data
A catalogue record for this book is available from the British Library

ISBN 978 1 846905 95 7

External project management by Gillian Lindsey
Edited by Geoff Amor
Typeset by Pantek Arts Ltd, Maidstone, Kent
Original illustrations © Pearson Education 2009
Illustrated by Pantek Arts Ltd, Maidstone, Kent
Cover photo © Shutterstock: William Attard McCarthy

Printed in Great Britain by Henry Ling Ltd, at the Dorset Press, Dorchester, Dorset

Acknowledgements
Where exam questions are taken from papers specified at the end of the question, these are reproduced by kind permission of Edexcel.

The publishers are grateful to Damian Riddle for writing the Answering multiple choice and extended questions, Anne Scott and Elizabeth Swinbank at University of York Science Education Group for writing the Revision techniques section and to Tim Tuggey and Andrea Gostick for their collaboration in reviewing this book.

Every effort has been made to contact copyright holders of material reproduced in this book.
Any omissions will be rectified in subsequent printings if notice is given to the publishers.

Disclaimer
This material has been published on behalf of Edexcel and offers high-quality support for the delivery of Edexcel qualifications.

This does not mean that the material is essential to achieve any Edexcel qualification, nor does it mean that it is the only suitable material available to support any Edexcel qualification. Edexcel material will not be used verbatim in setting any Edexcel examination or assessment. Any resource lists produced by Edexcel shall include this and other appropriate resources.

Copies of official specifications for all Edexcel qualifications may be found on the Edexcel website – www.edexcel.com

Contents

Unit 1 Physics on the go

Section 1 Mechanics

Section 2 Materials

Unit 2 Physics at work

Section 3 Waves

Section 4 DC electricity

Section 5 Nature of light

Unit 3 Exploring physics

Answers

How to use this Revision Guide

Welcome to your **Edexcel AS Physics Revision Guide**, perfect whether you're studying Salters Horners Advanced Physics (the blue book), or the 'concept-led' approach to Edexcel Physics (the red book).

This unique guide provides you with tailored support, written by Senior Examiners. They draw on real 'ResultsPlus' exam data from past A-level exams, and have used this to identify common pitfalls that have caught out other students, and areas on which to focus your revision. As you work your way through the topics, look out for the following features throughout the text:

ResultsPlus Examiner Tip
These sections help you perform to your best in the exams by highlighting key terms and information, analysing the questions you may be asked, and showing how to approach answering them. All of this is based on data from real-life A-level students!

ResultsPlus Watch Out!
The examiners have looked back at data from previous exams to find the common pitfalls and mistakes made by students – and guide you on how to avoid repeating them in *your* exam.

Quick Questions
Use these questions as a quick recap to test your knowledge as you progress.

Thinking Task
These sections provide further research or analysis tasks to develop your understanding and help you revise.

Worked Examples
The examiners provide step-by-step guidance on complex equations and concepts.

Each section also ends with:

Section Checklist
This summarises what you should know for this section, which specification point each checkpoint covers and where in the guide you can revise it. Use it to record your progress as you revise.

ResultsPlus Build Better Answers
Here you will find sample exam questions with exemplar answers, examiner tips and a commentary comparing both a basic and an excellent response: so you can see how to get the highest marks.

Practice exam questions
Exam-style questions, including multiple-choice, offer plenty of practice ahead of the written exams.

Both Unit 1 and Unit 2 conclude with a **Practice Unit Test** to test your learning. These are not intended as timed, full-length papers, but provide a range of exam-style practice questions covering the range of content likely to be encountered within the unit test.

The final Unit consists of advice and support on research, planning and analysis and evaluation skills for your assessed practical work, which is based on either a case study or a visit that involves an application of physics.

Answers to all the in-text questions, as well as detailed, mark-by-mark answers to the practice exam questions, can be found at the back of the book.

We hope you find this guide invaluable. Best of luck!

Revision techniques

Getting started can be the hardest part of revision, but don't leave it too late. Revise little and often! Don't spend too long on any one section, but revisit it several times, and if there is something you don't understand, ask your teacher for help. Just reading through your notes is not enough. Take an active approach using some of the revision techniques suggested below.

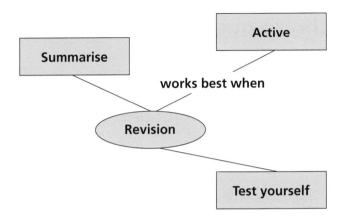

Summarising key ideas

Make sure you don't end up just copying out your notes in full. Use some of these techniques to produce condensed notes.

- Tables and lists to present information concisely.
- Index cards to record the most important points for each section.
- Flow charts to identify steps in a process.
- Diagrams to present information visually.
- Spider diagrams and concept maps to show the links between ideas.
- Mnemonics to help you remember lists.
- Glossaries to make sure you know clear definitions of key terms.

Include page references to your notes or textbook. Use colour and highlighting to pick out key terms.

Active techniques

Using a variety of approaches will prevent your revision becoming boring and will make more of the ideas stick. Here are some methods to try.

- Explain ideas to a partner and ask each other questions.
- Make a podcast and play it back to yourself.
- Use PowerPoint® to make interactive notes and tests.
- Search the Internet for animations, tests and tutorials that you can use.
- Work in a group to create and use games and quizzes.

Test yourself

Once you have revised a topic, you need to check that you can remember and apply what you have learnt.

- Use the questions from your textbook and this revision guide.
- Get someone to test you on key points.
- Try some past exam questions.

> If you use resources from elsewhere, make sure they cover the right content at the right level.

Section A of Unit tests 1 and 2 contains objective test (multiple-choice) questions. Section B contains a mixture of short-answer and extended-answer questions, including the analysis, interpretation and evaluation of experimental and investigative activities. In both sections you may be required to apply your knowledge and understanding of physics to situations that you have not seen before.

Multiple-choice questions

For each question there are four possible answers, labelled A, B, C and D. A good multiple-choice question (from an examiner's point of view) gives the correct answer and three other possible answers, which all seem plausible.

The best way to answer a multiple-choice question is to read the question and try and answer it *before* looking at the possible answers. You may need to do some calculations – space is provided on the question paper for rough working. If the answer you thought of or calculate is among the possible answers – job done! Just have a look at the other possibilities to convince yourself that you were right.

If the answer you thought of isn't there, look at the possible answers and try to eliminate wrong answers until you are left with the correct one.

You don't lose any marks by having a guess (if you can't work out the answer) – but you won't score anything by leaving the answer blank. If you narrow down the number of possible answers, the chances of having a lucky guess at the right answer will increase.

To indicate the correct answer, put a cross in the box following the correct statement. If you change your mind, put a line through the box and fill in your new answer with a cross.

How Science Works

The idea behind 'How Science Works' (HSW) is to give you insight into the ways in which scientists work: how an experiment is designed, how theories and models are put together, how data is analysed, how scientists respond to factors such as ethics and so on.

Many of the HSW criteria require practical or investigative skills and will be tested as part of your assessed practical work. However, there will be questions on the written units that cover all the HSW criteria. Some of these questions will involve data or graph interpretation, including the possible physical significance of the area between a curve and the horizontal axis, and determining quantities (with appropriate physical units) from the gradient and intercept of a graph (including at A2, log-linear and log-log graphs).

Another common type of HSW question will be on evaluating various steps in an experiment. For example,

- explain or justify why a particular piece of apparatus is used
- identify possible sources of systematic or random error
- explain why we use an instrument in a particular way
- what safety precautions would be relevant, and why?

You may be asked questions involving designing an investigation: these are likely to involve pieces of familiar practical work.

Other HSW questions may concentrate on issues surrounding the applications and implications of science (including ethical issues), or on using a scientific model to make predictions.

Extended questions

Remember that if part of a question is worth 6 marks, you need to make six credit-worthy points. Think about the points that you will make and put them together in a logical sequence when you write your answer. On longer questions, the examiners will be looking at your QWC (Quality of Written Communication) as well as the answer you give.

Motion equations and graphs

Scalar and vector quantities

Physical quantities (i.e. things you can measure or calculate) can be classified as either **scalars** or **vectors**. A scalar quantity only has magnitude, not direction, but with a vector quantity you need to state the direction as well as the magnitude. For example, a bag of sugar has a mass of 1.0 kg, but the force on it due to gravity is 9.8 N *downwards*.

Common scalar quantities: distance, speed, mass, volume, energy, power.
Common vector quantities: displacement, velocity, acceleration, force.

Equations of motion

There are several different equations used for calculations involving motion.

Quantities

s = displacement or distance

u = initial velocity or speed

v = final velocity or speed

a = acceleration

t = time taken

Equations for uniform acceleration

You will be given these equations of motion:

$v = u + at$

$s = ut + \frac{1}{2}at^2$

$v^2 = u^2 + 2as$

You need to be able to rearrange the equations to make any quantity the subject. A common rearrangement is $a = (v - u)/t$. This is just the definition of **acceleration**: acceleration = change in velocity/time.

These equations only apply where the acceleration is uniform (i.e. constant), which does include zero acceleration. Displacement, velocity and acceleration are all vector quantities, so you need to define one direction as positive and the opposite direction as negative.

Worked Example

A stone is thrown upwards at the edge of a cliff with an initial velocity of 25 m s^{-1}. Find its displacement after 7.0 s. Acceleration of free fall due to gravity, g = 9.81 m s^{-2}.

Decide that the upwards direction is positive and downwards is negative and list the *suvat* quantities:

s = to be found
u = +25 m s^{-1}
a = g = −9.81 m s^{-2}
t = 7.0 s

Decide on the required equation:

s, u, a and t are all in the equation $s = ut + \frac{1}{2}at^2$

Substitute the values and solve:

$s = ut + \frac{1}{2}at^2$
$s = 25 \text{ m s}^{-1} \times 7.0 \text{ s} + \frac{1}{2}(-9.81 \text{ m s}^{-2}) \times (7.0 \text{ s})^2$
$s = -65.3 \text{ m}$

This answer is negative, so it means that the stone is below the starting point.

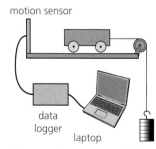

motion sensor

data logger

laptop

Using a motion sensor and data logger to graph motion

Making graphs of motion

Motion can be represented using graphs. Traditionally, data might have been collected using rulers and stopwatches. ICT methods are now readily available, for example using a motion sensor, as in the diagram on page 8.

This setup records displacement at regular intervals and can be used to create graphs of displacement against time or velocity against time. Light gates can also be used to measure motion. Both of these methods will eliminate human error, so time or displacement will be measured more *accurately*. They also allow greater *precision* than stopwatches. The data will have improved *validity* and *reliability*. With light gates, the length of the object interrupting the beam, or the distance between the gates, must still be measured manually.

Using graphs of motion

Displacement-time graphs

The line on a displacement-time graph is straight if the object is moving with constant velocity. A curve shows that the object is accelerating.

The gradient of a displacement-time graph is change in displacement/change in time, which is velocity,

$$v = \frac{\Delta s}{\Delta t}$$

This is also described as rate of change of displacement. Note that the units for the gradient are the units of the *y*-axis divided by the units of the *x*-axis, i.e. $\mathrm{m\,s^{-1}}$. If the object is accelerating, the velocity is found from the gradient of a tangent to the line.

Velocity-time graphs

The gradient of a velocity-time graph is the change in velocity/change in time, which is acceleration,

$$a = \frac{\Delta v}{\Delta t}$$

This is also described as rate of change of velocity. The area between the line and the time axis on a velocity-time graph is equal to the displacement.

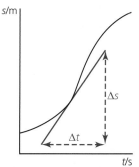

Displacement-time graph

Worked Example

Use the graph to calculate the displacement after 8 s.

Find the area of part A:

$\text{area of triangle} = \frac{1}{2} \times \text{base} \times \text{height}$

$\qquad\qquad\qquad = \frac{1}{2} \times 5\,\mathrm{s} \times 10\,\mathrm{m\,s^{-1}}$

$\qquad \text{displacement} = 25\,\mathrm{m}$

Find the area of part B:

$\text{area of rectangle} = \text{base} \times \text{height} = (8\,\mathrm{s} - 5\,\mathrm{s}) \times 10\,\mathrm{m\,s^{-1}}$
$\text{displacement} = 30\,\mathrm{m}$

Find displacement from the total area of A plus B:

$\text{displacement} = 25\,\mathrm{m} + 30\,\mathrm{m} = 55\,\mathrm{m}$

❓ Quick Questions

Q1 A swimming pool has a length of 50 m. Explain the difference between the *displacement* of a swimmer and the *distance* she has swum after completing three lengths.

Q2 A small rocket is launched upwards with a speed of $35\,\mathrm{m\,s^{-1}}$.
  a Calculate its speed and height after a time of 5.0 s.
  b Find the time taken to reach its maximum height.

Combining and resolving vectors

Combining vectors

Vector quantities have direction as well as magnitude, so when they are combined the *directions* must be taken into account. Mass is a scalar, so if we add masses of 3 kg and 4 kg, the total will always be 7 kg. But the combined effect of forces of 3 N and 4 N depends on their directions. In the diagrams, the magnitudes of the **resultant** forces (shown in red) are 7 N, 5 N and 1 N, respectively.

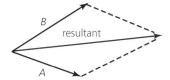

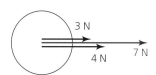

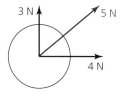

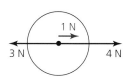

The sum of vectors depends on their directions

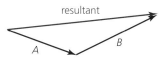

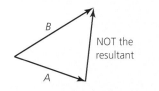

Vector parallelogram and vector triangle

You can find the resultant of two vectors using scale drawings or by calculation. Take two forces, *A* and *B*, acting on a body. The two forces can be shown as two sides of a parallelogram, as in the upper diagram. The diagonal of the parallelogram is the resultant.

It is often more convenient to draw a triangle, as in the lower diagram. The two vectors are drawn 'tip-to-tail' and the resultant is the third side. Note that it goes from the tail of *A* to the tip of *B*.

Combining perpendicular vectors

When two vectors are at right angles to each other, it may be easier to find the resultant by calculation, using Pythagoras' theorem and trigonometry.

Worked Example

A model aeroplane flies northwards with velocity 15 m s⁻¹ while a 9 m s⁻¹ wind is blowing eastwards. Find the velocity of the plane over the ground.

First sketch the diagram. Then use Pythagoras' theorem:

$$(\text{resultant velocity})^2 = (15\,\text{m s}^{-1})^2 + (9\,\text{m s}^{-1})^2$$
$$(\text{resultant velocity})^2 = 306\,\text{m}^2\,\text{s}^{-2}$$
$$\text{resultant velocity} = \sqrt{306} = 17.5\,\text{m s}^{-1}$$

To find the angle, use the tangent:

$$\tan\theta = \frac{9\,\text{m s}^{-1}}{15\,\text{m s}^{-1}}$$
$$\theta = 31°$$

So the velocity of the aeroplane is 17.5 m s⁻¹ at an angle of 31° from the north.

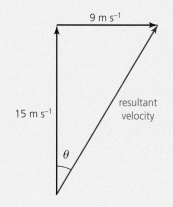

Resolving vectors

A single vector can be represented as the sum of two perpendicular vectors, known as its **components**. Vectors at right angles can then be treated independently, and can have their magnitudes added to other parallel vectors if necessary.

The components of a vector are found by using that vector as the diagonal of a parallelogram of vectors or as the hypotenuse of a right-angled triangle of vectors.

Worked Example

A guy rope pulls on a tent peg with a force of 30N at an angle of 35° to the horizontal. Calculate the horizontal and vertical components of the force.

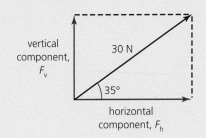 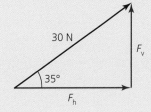

First sketch the diagram. Then find the horizontal component, using the cosine:

$$\cos 35° = \frac{F_h}{30\,N}$$
$$F_h = 30\,N \times \cos 35° = 24.6\,N$$

To find the vertical component, use the sine:

$$\sin 35° = \frac{F_v}{30\,N}$$
$$F_v = 30\,N \times \sin 35° = 17.2\,N$$

Combining more than two vectors

If there are more than two vectors, they can be combined by continuing the *tip-to-tail* procedure. The resultant is found by drawing from the tail of the first vector to the tip of the last.

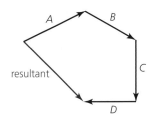

Combining several vectors

? Quick Questions

Q1 A boat pointing due north across a river travels through the water at $10\,m\,s^{-1}$. The river is flowing at $6\,m\,s^{-1}$ due east.
  a Use a scale drawing to find the resultant velocity of the boat.
  b Check your answer by calculating the resultant velocity.
Q2 A ball is kicked with velocity of $8.0\,m\,s^{-1}$ at an angle of 38° to the ground. Calculate the vertical and horizontal components of its velocity.

⚙ Thinking Task

A school treasure hunt uses the following set of directions: A, 80m at 045°; B, 40m at 150°; C, 50m at 180°; D, 60m at 270°.
a Team 1 follows the directions correctly. Draw a vector diagram to find their final displacement.
b Team 2 gets the directions mixed up and follows them in the order C–B–D–A. What effect will this have on their final position compared to Team 1?

Force and acceleration

Newton's first law of motion

This states says that, if the forces acting on a body are in **equilibrium**, its velocity will remain constant.

> ### Worked Example
>
> A submarine is moving forwards at a constant speed of 15 m s⁻¹ and is ascending (moving upwards) at a steady rate of 2 m s⁻¹. Choose the option that correctly compares the magnitudes of the forces:
>
> **A** upthrust > weight; thrust > drag
> **B** upthrust > weight; thrust = drag
> **C** upthrust = weight; thrust > drag
> **D** upthrust = weight; thrust = drag
>
>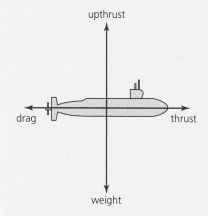
>
> The correct answer is **D**. The horizontal and vertical components of velocity are constant, and therefore the forces are in equilibrium.

Newton's second law of motion

If there is a resultant force, then there will be a change in velocity, i.e. an acceleration.

Newton's second law states that the acceleration of an object is proportional to the resultant force and inversely proportional to its mass. Taking these together:

$$\text{resultant force} = \text{mass} \times \text{acceleration}$$

or

$$\sum F = ma$$

where F is the force in N, m is mass in kg, and a is acceleration in m s⁻² ($\sum F$ means the sum of all the forces, i.e. the resultant force). Acceleration and force are both vectors and must act in the same direction.

If the resultant force is zero, then acceleration will be zero. So Newton's first law is just a special case of the second law: if $\sum F = 0$, then $\Delta v = 0$.

A motorcycle with rider has a mass of 520 kg and accelerates at $3.5\,\text{m s}^{-2}$.

a If the drag is 430 N, calculate the forward driving force on the motorcycle.

b A passenger of mass 70 kg gets on the motorcycle. What will the initial acceleration be now?

- -

a Find the resultant force, and then add the drag to find the driving force:

resultant force = mass × acceleration = $520\,\text{kg} \times 3.5\,\text{m s}^{-2}$ = 1820 N

total forwards force = 1820 N + 430 N = 2250 N

b Assume there is no drag initially:

$$\text{acceleration} = \frac{\text{force}}{\text{mass}} = \frac{2250\,\text{N}}{590\,\text{kg}} = 3.8\,\text{m s}^{-2}$$

Newton's third law of motion

This states that, if body A exerts a force on body B, then body B exerts a force of the same type on body A that is equal in magnitude and opposite in direction.

To describe the Newton's third law pair force in a given situation, just keep the magnitude and type of force the same and swap the direction and body acted on. For example, the Earth exerts a downwards gravitational force of 10 N on a rock, so the rock exerts an upwards gravitational force of 10 N on the Earth.

The diagram shows three forces of equal magnitude relating to a box on a table. Explain why only one pair of equal and opposite forces make a Newton's third law pair.

- -

Forces A and C are equal in magnitude and opposite in direction. They are both normal contact forces, i.e. the same type. Force A acts on the table and C acts on the box, i.e. different bodies. So forces A and C make a Newton's third law pair.

Forces B and C are equal in magnitude and opposite in direction. Force B is a gravitational force and C is a normal contact force, i.e. different types. Forces B and C both act on the box, i.e. the same body. So forces B and C do not make a Newton's third law pair – they are forces in equilibrium acting on the box.

ResultsPlus
Watch out!

In questions about Newton's third law, students often go wrong because they aren't clear about which body is exerting which force.

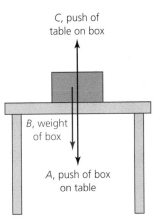

C, push of table on box

B, weight of box

A, push of box on table

ResultsPlus
Watch out!

Everyone remembers that the forces in a Newton's third law pair are 'equal and opposite'. The bit that students find tricky is remembering that they act on *different* bodies. In this example, forces B and C would relate better to Newton's first law.

Quick Questions

Q1 A passenger of mass 70 kg is in a lift accelerating upwards at a rate of $1.5\,\text{m s}^{-2}$. What force does he feel on his feet due to this acceleration?

Q2 A cyclist and bicycle of combined mass 109 kg are travelling up a hill at a steady speed of $15\,\text{m s}^{-1}$, but using the maximum available force. With the force removed, they come to rest in 3.5 s.
 a Calculate the decelerating force.
 b Assuming the forces remain the same, what is the maximum speed which could be attained by cycling downhill from rest for 6.5 s?

Q3 Explain why a skydiver has two different terminal velocities at different stages during her jump.

Thinking Task

Make a table to summarise the differences and similarities between a Newton's third law pair of forces and two forces in equilibrium.

Gravity and free-body diagrams

The Earth has a gravitational field that extends into space. The **gravitational field strength** g at the Earth's surface is $9.81\,\text{N kg}^{-1}$. That means that a body of mass 1 kg will experience a downwards gravitational force, or **weight**, of 9.81 N:

$$\text{weight} = \text{mass} \times \text{gravitational field strength}$$

or $$W = mg$$

The gravitational field strength depends on position relative to other objects with mass, such as planets, and the general expression is:

$$\text{gravitational field strength} = \text{gravitational force/mass}$$

or $$g = \frac{F}{m}$$

Worked Example

An astronaut weighs 134 N while she is training on the Moon. On the Moon, $g = 1.61\,\text{N kg}^{-1}$.

a Calculate her mass.

b What will her weight be on Mars, where $g = 3.73\,\text{N kg}^{-1}$?

- -

a $g = F/m$

so $m = F/g = 134\,\text{N}/1.61\,\text{N kg}^{-1} = 83.2\,\text{kg}$

b $F = mg$

$= 83.2\,\text{kg} \times 3.73\,\text{N kg}^{-1} = 310\,\text{N}$

Acceleration of free fall

For a body of mass m, weight = mass × gravitational field strength, i.e. $W = mg$. If it is allowed to accelerate freely under this force, acceleration = force/mass, so $a = mg/m$. Therefore, $a = g$. In other words, all objects falling freely under gravity have the same acceleration, and the magnitude of this acceleration is the same as the magnitude of the gravitational field strength.

Centre of gravity

If you have ever tried to help lift a piano, you will know that gravity acts on all parts of an object. We can simplify the situation when we are calculating forces by assuming that *all* the gravitational force acts on a single point, called the **centre of gravity**. Very often the whole object is then treated as a single point mass.

• centre of gravity

Centre of gravity of some regular objects

Free-body force diagrams

It is important to be able to distinguish between the forces acting *on* a body and the forces exerted *by* the body. A **free-body** force diagram shows *all* the forces acting *on* the body, and none of the forces it is exerting on other objects.

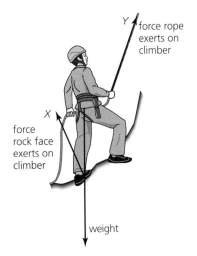

Forces on a rock climber

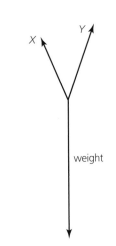

Forces on the rock climber, treated as a point object

The rock climber is exerting forces on the rope and the ground, but these are not included in the free-body force diagram for the climber – only the forces acting *on* the climber are shown. In this case, the lines of action of all the forces pass through a single point, and it is easier to show them as acting on a point object before carrying out calculations.

Forces in equilibrium

Once you have a free-body diagram, you can combine the forces with a vector diagram. If this produces a closed vector polygon, the resultant is zero. This means they are in **equilibrium**. For example, if we draw a vector diagram for the climber, we see that the three forces from the free-body diagram are in equilibrium.

If you know the forces are in equilibrium, you can complete a closed vector polygon. By scale drawing or trigonometry, any unknown forces can be calculated.

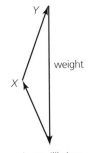

Forces in equilibrium

Quick Questions

Q1 On a particular planet, a mass of 8 kg has a weight of 28 N. On another planet with twice the gravitational field strength, what is the weight of a 16 kg mass?

  A 14 N **B** 28 N **C** 56 N **D** 112 N

Q2 The diagram shows a block sliding down a plank at a constant speed.
  a What can you say about the forces?
  b The weight of the block is 5 N and the plank is at an angle of 35° to the horizontal. Sketch a vector diagram for the three forces and use it to calculate the friction.

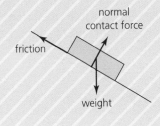

Projectile motion

We can find a resultant velocity from the vector sum of two separate velocities, such as a boat moving across a flowing river. As long as two velocities are perpendicular, the components can be treated entirely separately throughout the motion.

The vertical motion of all projectiles is subject to the acceleration of free fall. But, in the absence of other forces, the horizontal velocity remains constant.

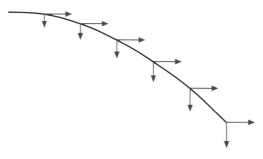

horizontal component v_h

vertical component v_v

resultant velocity

θ

Horizontal launch

The diagram shows an object that has been projected horizontally and is now falling freely under gravity. The horizontal velocity is the same at each point, but the downward velocity is increasing. The path followed is determined by the resultant velocity at each instant and is parabolic.

Horizontal and vertical velocities at successive time intervals

ResultsPlus
Examiner tip

Decide which vertical direction is going to be positive and apply it consistently in all projectile problems.

ResultsPlus
Examiner tip

In order to avoid awkward quadratic equations, the initial height is generally the same as the final height in exam questions. If that is the case, you can sometimes save time by assuming that the flight is completely symmetrical, i.e. initial vertical velocity = − final vertical velocity, and time to fall back to initial height = 2 × time to reach top of flight.

Worked Example

A ball is kicked off a flat roof with a horizontal velocity of $12\,\mathrm{m\,s^{-1}}$. How would you:

a calculate the horizontal distance travelled after it has fallen 15 m?

b calculate the resultant velocity after 1.9 s?

- -

a Consider vertical motion only and work out the time it would take for the object to reach the ground, accelerating under gravity (refer to the equations of motion on page 8). Then use this time and the horizontal velocity to work out the distance it will travel in that time.

b Find the vertical component of velocity at 1.9 s (remembering that $a = g = 9.81\,\mathrm{m\,s^{-2}}$). Then combine the horizontal and vertical components to find the magnitude (by Pythagoras' theorem) and angle (using trigonometry).

Launch at other angles

If an object is not launched horizontally or vertically, you start a problem by resolving the initial velocity into horizontal and vertical components. A problem will often involve calculating the time of flight by considering the vertical motion. Find the time to reach the maximum height (when the vertical component of velocity is zero) and double it. You then find the maximum horizontal distance travelled (known as the range) from the time of flight and the horizontal velocity.

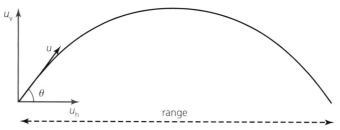

u_v

u

θ

u_h

range

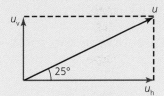

Worked Example

A ball is kicked at a velocity of $11\,\text{m s}^{-1}$ at an angle of $25°$ to the horizontal. Calculate the distance it travels before reaching the ground.

Considering vertical motion, first find the vertical component of initial velocity:

$u_v = 11\,\text{m s}^{-1} \times \sin 25° = 4.6\,\text{m s}^{-1}$

Find the time at which $v = 0\,\text{m s}^{-1}$. You have:

$u = 4.6\,\text{m s}^{-1}$, $a = g = -9.81\,\text{m s}^{-2}$, $v = 0\,\text{m s}^{-1}$, $t = ?$

Use $v = u + at$, and rearrange:

$at = v - u$

$t = \dfrac{v - u}{a}$

$ = \dfrac{0 - 4.6\,\text{m s}^{-1}}{-9.81\,\text{m s}^{-2}}$

$t = 0.47\,\text{s}$

Total time of flight $= 2 \times 0.47\,\text{s} = 0.94\,\text{s}$

Find the horizontal component of initial velocity:

$u_h = 11\,\text{m s}^{-1} \times \cos 25° = 10\,\text{m s}^{-1}$

Now you have: $v = 10\,\text{m s}^{-1}$, $t = 0.94\,\text{s}$, $s = ?$

Find s using $s = vt$ and substituting:

$s = 10\,\text{m s}^{-1} \times 0.94\,\text{s} = 9.4\,\text{m}$

ResultsPlus
Watch out!

Questions often ask for an explanation of whether this is likely to be the true distance. Students often just say 'No, because of air resistance'. But you are expected to state whether the distance will be more or less than your answer, and to explain why air resistance has this effect, e.g. by causing a deceleration, reducing both its speed and the time the projectile is in the air.

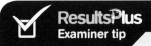

Quick Questions

Q1 During a school production of *Macbeth*, a dagger is kicked and slides off the stage with an initial horizontal velocity of $1.5\,\text{m s}^{-1}$. It lands on the floor a horizontal distance of $80\,\text{cm}$ away. Calculate the height of the stage.

Q2 Water from a garden hose at ground level is squirted into the air at an angle of $65°$ to the horizontal. The water reaches a height of $2.5\,\text{m}$ above the hose.
  a Sketch the path of the water.
  b By considering vertical motion only, calculate the vertical component of the water's initial velocity.
  c Use the vertical component to find the horizontal component of velocity.

Q3 Find the distance travelled by a ball kicked from ground level at a speed of $12\,\text{m s}^{-1}$ and an angle of $33°$ to the horizontal.

ResultsPlus
Examiner tip

After completing a problem involving everyday situations, e.g. a school stage, think about your answer to see if it is reasonable. If it isn't, check through your working again. (You did already check it once, didn't you?)

Thinking Task

In the absence of air resistance, the maximum range is achieved for a launch angle of $45°$ to the horizontal. By thinking about the effect of air resistance on vertically upward motion, vertically downward motion and horizontal motion, draw and describe the path of a real projectile.

Work and power

When a force acts on an object and transfers energy, **work** has been done. For example, suppose someone pushes a car, which accelerates and gains kinetic energy. Energy is transferred *from* the person *to* the car as work is done *by* the person *on* the car.

work done = force × displacement in the direction of the force

$$\Delta W = F\Delta s$$

where work W is in joules, force F is in newtons and distance s is in metres. The symbol Δ stands for a change, so think of this as

change in energy = force × change in position

Worked Example

A force of 520 N is used to push a car through a distance of 19 m.

a Show that the work done by the person pushing the car is about 10 000 J.

b The kinetic energy transferred is found to be only 8000 J. Suggest a reason why.

- -

a $\Delta W = F\Delta s = 520\,\text{N} \times 19\,\text{m} = 9.9 \times 10^3\,\text{J}$.

b There are resistive forces, so the resultant force acting on the car is less than 520 N. Some of the work done on the car has been transferred to heat energy, rather than increasing the car's kinetic energy.

When the force does work with no opposition, energy is transferred to kinetic energy. However, work is usually done against some opposing force, such as friction or drag, in which case part of the energy will be transferred as heat. When work is done against a gravitational force, the transfer is to gravitational potential energy. When you stretch an elastic material, the chief transfer is to elastic potential (strain) energy, although you need to watch out because some can be transferred to heat as well.

Although work done is a scalar quantity, force and displacement are vectors, so their directions important. If force and displacement are not in the same direction, use the component of force in the direction of the displacement (or the component of the displacement in the direction of the force).

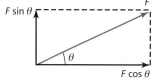

Finding the component of the force in the direction of the displacement

In the diagram, if the displacement is horizontal, the component of the force in the direction of the displacement is $F\cos\theta$. So, work done = $F\cos\theta \times s$.

A lawnmower handle is at an angle of 43° to the horizontal. A gardener pushes along the handle with a force of 150 N. How much work is done by the gardener in pushing the lawnmower a distance of 25 m?

We have work done = $\Delta W = F\Delta s$ and horizontal component of force = $F\cos\theta$. So we find:

$$\text{work done} = F\cos\theta \times s = 150\,\text{N} \times \cos 43° \times 25\,\text{m} = 2 \times 10^3\,\text{J}$$

ResultsPlus
Watch out!

Be sure you know your calculator. If I press the keys in the order 1, 5, 0, ×, cos, 4, 3, ×, 2, 5, =, I get the answer 149 because it works out the cosine of (43 × 25). If you put the cosine calculation last it's OK.

Power

Power is the rate at which work is done, which is the same as saying that power is the rate at which energy is transferred. Power can be calculated using this formula:

$$\text{power} = \frac{\text{work done}}{\text{time}}$$

or

$$P = \frac{W}{t}$$

Similarly,

$$\text{power} = \frac{\text{energy transferred}}{\text{time}}$$

where work (or energy) W is measured in joules, time t is in seconds and power P is in joules per second, or watts.

A person stacking shelves lifts crates of weight 70 N through a height of 1.5 m. If it takes two minutes to lift 20 crates, what is the rate at which work is done?

Find the work done lifting one crate. Work is done against gravitational force, so F is the weight:

$$\Delta W = F\Delta s = 70\,\text{N} \times 1.5\,\text{m} = 105\,\text{J}$$

$$\text{power} = \frac{\text{total work done}}{\text{time taken}} = \frac{105\,\text{J} \times 20}{120\,\text{s}} = 17.5\,\text{W}$$

ResultsPlus
Watch out!

Time is often quoted in units other than seconds, so be sure you convert to seconds when doing power calculations. Here two minutes is 120 s.

⟨?⟩ Quick Questions

Q1 For modern travellers, wheeled suitcases with retractable handles have become more popular than backpacks. A traveller pulls such a case using a force of 65 N at an angle of 50° to the horizontal.
  a Calculate the work done pulling it a distance of 85 m through the airport.
  b It takes the traveller one minute to cover the 85 m. What power is she exerting?

Q2 An escalator has 50 steps available at any time, each capable of holding one passenger of weight 700 N. It lifts the passengers between floors separated vertically by 12.5 m. The rate of vertical climb is 0.3 m s⁻¹.
  a Find the time for a passenger to reach the next floor.
  b Calculate the maximum rate at which the escalator transfers gravitational potential energy to the passengers.
  c Explain three reasons why the power required by the motor would be greater than this.

Energy and energy conservation

Gravitational potential energy

When an object is lifted, work is done against the downward gravitational force, i.e. the weight. The distance moved is the vertical height through which the object is raised. Therefore,

$$\text{change in gravitational potential energy} = \text{weight} \times \text{change in height}$$

$$\Delta E_{grav} = W \times \Delta h$$

and, as weight = mass × g:

$$\Delta E_{grav} = mg\Delta h$$

The *change* in **gravitational potential energy** is related to the *change* in height – you don't need the total height above the ground. Also note that, although this is related to a specific direction, it is still a scalar quantity.

> ### Worked Example
>
> An aeroplane of mass 1400 kg climbs from an altitude of 1800 m to 3500 m. Calculate its change in gravitational potential energy.
>
> $\Delta E_{grav} = mg\Delta h = 1400\,\text{kg} \times 9.81\,\text{N}\,\text{kg}^{-1} \times (3500\,\text{m} - 1800\,\text{m}) = 2.3 \times 10^7\,\text{J}$

Kinetic energy

Moving bodies have kinetic energy by virtue of their motion. The formula is:

$$\text{kinetic energy} = \tfrac{1}{2} \times \text{mass} \times (\text{speed})^2$$

$$E_k = \tfrac{1}{2}mv^2$$

A change in kinetic energy can also be found based on work done.

> ### Worked Example
>
> An aeroplane is travelling with a speed of 45 m s^{-1}. Calculate its kinetic energy.
>
> $E_k = \tfrac{1}{2}mv^2 = \tfrac{1}{2} \times 1400\,\text{kg} \times (45\,\text{m}\,\text{s}^{-1})^2 = 1.4 \times 10^6\,\text{J}$

Conservation of energy

Energy can be neither created nor destroyed, but it can be transferred. This means that, in any process, total energy at the start = total energy at the end. This principle is often used as part of an exam question. A particularly common example is the transfer of kinetic energy to gravitational potential energy or vice versa.

Worked Example

A child of mass 22 kg sits on a swing and is pulled back, raising her through a height of 1.2 m. Calculate her speed as she swings through the lowest point.

First find the change in gravitational potential energy:

$$\Delta E_{grav} = mg\Delta h = 22\,\text{kg} \times 9.81\,\text{N kg}^{-1} \times 1.2\,\text{m} = 259\,\text{J}$$

(Notice that the distance she is pulled back isn't required – just her change in height.)

Now assume that *all* the gravitational potential energy is transferred to kinetic energy:

$$E_k = \frac{1}{2}mv^2$$

so

$$v^2 = \frac{2E_k}{m}$$

$$v = \sqrt{\frac{2 \times 259\,\text{J}}{22\,\text{kg}}} = 4.9\,\text{m s}^{-1}$$

ResultsPlus
Examiner tip

If you take the equations for gravitational potential energy and kinetic energy together for a transfer like this, then $mg\Delta h = \frac{1}{2}mv^2$. The mass m is the same, so it cancels out. This gives $g\Delta h = \frac{1}{2}v^2$, which can save some time.

Another common transfer involves the transfer from or to elastic strain energy. In this case use $\Delta E_{el} = \frac{1}{2}F\Delta x$ if the extension obeys Hooke's law (see page 30). Otherwise you may have to find the energy from the area under a graph of extension against force.

Quick Questions

Q1 Newton's cradle has five steel balls suspended in a frame. When the end one is raised and released, it strikes the rest with a sharp click. The energy is transferred through the balls, and the ball at the opposite end then moves away to about the same height as the first one was dropped from, as shown. The process repeats in reverse and goes on until the motion gradually dies away.

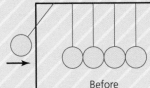

Before   After

Newton's cradle

a The balls each have a mass of 0.080 kg. If the first ball is raised through a height of 1.5 cm, calculate its speed just before it strikes the other balls.
b Explain as fully as possible why the motion dies away.
Q2 A car of mass 1200 kg is travelling at a speed of 25 m s^{-1}.
a What braking force must be applied to stop it in a distance of 120 m? Calculate your answer using kinetic energy and work done by the braking force.
b Now calculate your answer using equations of motion and Newton's second law.
c Which method seems more straightforward?

Thinking Task

Explain why a knowledge of different forms of energy is important for an engineer designing sports safety equipment, such as cricket pads or climbing helmets.

Mechanics checklist

By the end of this section you should be able to:

Revision spread	Checkpoints	Spec. point	Revised	Practice exam questions
Motion equations and graphs	Use the equations for uniformly accelerated motion in one dimension: $v = u + at$ $s = ut + \frac{1}{2}at^2$ $v^2 = u^2 + 2as$	1	☐	☐
	Demonstrate an understanding of how ICT can be used to collect data for, and display, displacement-time and velocity-time graphs for uniformly accelerated motion and compare this with traditional methods in terms of reliability and validity of data	2	☐	☐
	Identify and use the physical quantities derived from the slopes and areas of displacement-time and velocity-time graphs, including cases of non-uniform acceleration	3	☐	☐
	Distinguish between scalar and vector quantities and give examples of each	5	☐	☐
Combining and resolving vectors	Resolve a vector into two components at right angles to each other by drawing and by calculation	6	☐	☐
	Combine two coplanar vectors at any angle to each other by drawing, and at right angles to each other by calculation	7	☐	☐
	Draw and interpret free-body force diagrams to represent forces on a particle	8	☐	☐
Force and acceleration	Use $\sum F = ma$ in situations where m is constant (Newton's first law of motion ($a = 0$) and second law of motion)	9	☐	☐
	Identify pairs of forces constituting an interaction between two bodies (Newton's third law of motion)	11	☐	☐
Gravity and free-body diagrams	Draw and interpret free-body force diagrams to represent forces on a particle or on an extended but rigid body, using the concept of centre of gravity of an extended body	8	☐	☐
	Use the expressions for gravitational field strength $g = F/m$ and weight $W = mg$	10	☐	☐
Projectile motion	Recognise and make use of the independence of vertical and horizontal motion of a projectile moving freely under gravity	4	☐	☐
	Use the equations for uniformly accelerated motion in one dimension: $v = u + at$ $s = ut + \frac{1}{2}at^2$ $v^2 = u^2 + 2as$	1	☐	☐
Work and power	Use the expression for work $\Delta W = F\Delta s$, including calculations when the force is not along the line of motion	15	☐	☐
	Calculate power from the rate at which work is done or energy transferred	17	☐	☐
Energy and energy conservation	Apply the principle of conservation of energy, including use of work done, gravitational potential energy and kinetic energy	14	☐	☐
	Use the relationship $E_k = \frac{1}{2}mv^2$ for the kinetic energy of a body	12	☐	☐
	Use the relationship $\Delta E_{grav} = mg\Delta h$ for the gravitational potential energy transferred near the Earth's surface	13	☐	☐
	Understand some applications of mechanics, for example to safety or to sports	16	☐	☐

ResultsPlus
Build Better Answers

1. A student is using a trolley and track to investigate the relationship between force and acceleration.

a Once he has obtained data for force and acceleration, how could he demonstrate the expected relationship between the force and the acceleration? **[2]**

☑ Examiner tip

Many students fail to gain full marks for this kind of question because they do not include enough detail in their answers.

Student answer	Examiner comments
Plot a graph of force against acceleration.	This answer would get one of the two marks available. You also need to say that the graph should be a straight line *through the origin*, because it is expected that acceleration is directly proportional to force.

b In such an experiment, the track is given a slight tilt to compensate for friction. Why is this necessary if the relationship suggested by Newton's second law is to be successfully demonstrated? **[2]**

Adapted from Edexcel June 2007 Unit Test 1

Student answer	Examiner comments
Newton's law refers to the resultant force on an object. If the student does not compensate for friction, the resultant force accelerating the trolley will be less than the applied force.	This is an excellent answer. It is always worth looking at the number of marks available for a written question like this – if there are two marks, your answer needs to include two separate points.

2. The force produced by the engine of a car which drives it is ultimately transmitted to the area of contact between the car's tyres and the road surface. The diagram shows a wheel at an instant during the motion of the car when it is being driven forward in the direction indicated.

Two horizontal forces act at the point of contact between the tyre and road due to the transmitted force from the engine. These are shown as F_1 and F_2. Assume that the area of contact between the tyre and road is very small.

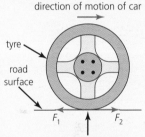

a Copy and complete these statements:
  (i) F_1 is the force of the .... on the ....
  (ii) F_2 is the force of the .... on the .... **[2]**

Student answer	Examiner comments
i F_1 is the force of the **tyre** on the **road**. ii F_2 is the force of the **friction** on the **car**.	Part **i** is correct, but part **ii** is not. These forces are a 'Newton's third law pair', equal in size and opposite in direction. The forces are caused by friction, but act on different objects. F_2 is the force of the road on the tyre.

b The total forward force on a car is 400 N when the car is travelling at a constant speed of 10 m s⁻¹ along a level road. Although work is done on the car, it continues to move at a constant speed. Explain why the car is not gaining kinetic energy. **[2]**

Adapted from Edexcel June 2007 Unit Test 1

Student answer	Examiner comments
The work done on the car is being transferred to thermal energy, because of friction.	This answer would get both marks. Be careful though – the friction is within the car itself (bearings, axles, etc.) and the drag force due to air resistance, *not* the friction between the tyres and the road. If you had mentioned the tyres in this answer, you would have lost one of the marks. You could also answer this question by stating that the forward force on the car is balancing air resistance/friction forces, so there is no change in speed.

Practice exam questions

1 Which of these quantities is *not* a vector quantity?

     A   acceleration
     B   kinetic energy
     C   displacement
     D   force

2 A boy is pulling a sledge. The rope is at an angle of 30° to the horizontal, and he is pulling with a force of 100 N. What is the horizontal component of the force?

     A   50 N
     B   58 N
     C   87 N
     D   93 N

3 Which statement best describes the location of the centre of gravity of a mug?

     A   In the centre of the space inside the mug.
     B   In the centre of the circular base.
     C   In the handle.
     D   Near the centre of the space inside the mug.

4 Jack carries a 20 kg suitcase up a flight of stairs in 5 s. Jill carries a 12 kg suitcase up the same stairs in 3 s. Which statement is correct?

     A   Jack's power was greater than Jill's.
     B   Jack and Jill both produced the same power.
     C   Jill's power was greater than Jack's.
     D   You need to know the height of the stairs to compare their power.

5 A hot-air balloon is rising vertically at a speed of $10\,\text{m s}^{-1}$. An object is released from the balloon. The graph shows how the velocity of the object varies with time from when it leaves the balloon to when it reaches the ground four seconds later. It is assumed that the air resistance is negligible.

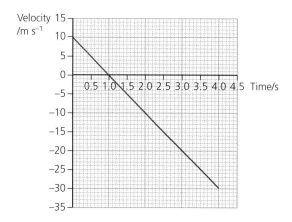

     a   Use the graph to
         i   show that the object continues to rise for a further 5 m after it is released. **[1]**
         ii   determine the total distance travelled by the object from when it is released from the balloon to when it reaches the ground. **[2]**
     b   Hence determine the object's final displacement from its point of release from the balloon. **[2]**
     c   Sketch a graph of acceleration (*y*-axis) against time showing how the acceleration of the object changes during the time from when it leaves the balloon to when it hits the ground. Mark any significant values on the axes. **[3]**

Adapted from *Edexcel January 2008 Unit Test 1*

6 a On 20 July 1969, Neil Armstrong became the first human to step onto the surface of the Moon. One experiment carried out was to film a hammer and a feather dropped at the same time from rest. They fell side by side all the way to the ground, falling a distance of 1.35 m in a time of 1.25 s.
  i Show that this gives a value for their acceleration of about 1.7 m s^{-2}. **[2]**
  ii Explain why the acceleration of both the hammer and the feather was constant throughout their fall. **[2]**

  In the questions that follow, assume that on the Moon:

  acceleration of free fall g = 1.7 m s^{-2}

  gravitational field strength g = 1.7 N kg^{-1}

b Neil Armstrong and his spacesuit had a total mass of 105 kg. Calculate his weight, including the spacesuit, on the Moon. **[2]**
c In 1971 another astronaut, Alan Shepherd, hit a golf ball on the Moon. He later said that it went '… miles and miles …'.
  i Suppose that he hit the ball so that it left his club at a speed of 45 m s^{-1}, at an angle of 20° to the horizontal. Calculate the time of flight of the ball. **[3]**
  ii Calculate the horizontal distance travelled by the ball before landing. **[2]**
  iii Comment on this distance in relation to his statement. 1 mile = 1.6 km **[1]**

Edexcel June 2007 PSA1

7 A weightlifter raised a bar of mass of 110 kg through a height of 2.22 m. The bar was then dropped and fell freely to the floor.

a Show that the work done in raising the bar was about 2400 J. **[2]**
b It took 3.0 s to raise the bar. Calculate the average power used. **[2]**
c State the principle of conservation of energy. **[2]**
d Describe how the principle of conservation of energy applies to
  i lifting the bar,
  ii the bar falling to the floor. Do not include the impact with the floor. **[3]**
e Calculate the speed of the bar at the instant it reaches the floor. **[3]**

Edexcel June 2006 Unit Test 1

Fluids and fluid flow 1

The term 'fluid' refers to a substance that can flow. That includes both liquids and gases.

Density

The definition of **density** is the mass per unit volume:

$$\rho = \frac{m}{V}$$

where mass m is in kg, volume V is in m^3 and density ρ is in $kg\,m^{-3}$.

In questions involving density, you may be asked to find the mass or the volume. You are expected to be able to find the volume of rectangular blocks, cylinders and spheres.

Upthrust

If a body is fully or partially submerged in a liquid, pressure differences at different depths cause it to experience an upward force known as **upthrust**. By Archimedes' principle, the upthrust is equal in magnitude to the weight of fluid displaced by the body. If the upthrust on a fully submerged object is less than its weight the object will sink. An object will float if it can displace its own weight of fluid without becoming fully submerged.

ResultsPlus
Watch out!

A common error made by students is to forget to halve the diameter when calculating the volume of a cylinder or a sphere.

If unit conversions are required (e.g. from cm to m), it is a good idea to do them before finding the volume. Students in a hurry sometimes forget that they have cubed the length dimension and state that $100\,cm^3 = 1\,m^3$ (instead of $1\,000\,000\,cm^3 = 1\,m^3$!).

Worked Example

A steel sphere of diameter 2.7 cm is immersed in oil of density $900\,kg\,m^{-3}$. Find the resultant force on the sphere at the moment it is released. Density of steel = $8100\,kg\,m^{-3}$.

- -

Find the volume of the body:

$$\text{volume of sphere} = \frac{4}{3}\pi r^3 = \frac{4}{3} \times \pi \times \left(\frac{0.027\,m}{2}\right)^3$$
$$= 1.0 \times 10^{-5}\,m^3$$

Find the mass of that volume of the surrounding fluid:

$$\text{mass of displaced fluid} = \text{density} \times \text{volume}$$
$$= 900\,kg\,m^{-3} \times 1.0 \times 10^{-5}\,m^3 = 9.0 \times 10^{-3}\,kg$$

Find the upthrust:

$$\text{upthrust} = \text{weight of displaced fluid} = \text{mass} \times g$$
$$= 9.0 \times 10^{-3}\,kg \times 9.81\,N\,kg^{-1}$$
$$= 0.088\,N \text{ (upwards)}$$

Find the resultant force:

$$\text{weight of ball} = \text{density} \times \text{volume} \times g$$
$$= 8100\,kg\,m^{-3} \times 1.0 \times 10^{-5}\,m^3 \times 9.81\,N\,kg^{-1}$$
$$= 0.795\,N \text{ (downwards)}$$
$$\text{resultant force} = 0.795\,N - 0.088\,N = 0.71\,N$$

ResultsPlus
Examiner tip

If you see something like 'at the moment it is released', it generally means that the velocity is zero, and therefore that there are no resistive forces.

ResultsPlus
Watch out!

Students often forget to multiply by g and use the result for mass as the upthrust.

Laminar and turbulent flow

Fluids moving through pipes or around obstacles can flow in different ways. You may be asked to identify, draw and describe both types of flow.

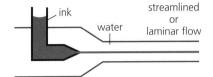

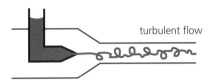

Laminar flow (or **streamline flow**) usually occurs at lower speeds and around more streamlined objects. Laminar flow means flow in layers. The layers do not mix, except on the molecular scale, so your drawings should not show streamlines crossing over. The layers are roughly parallel. The speed and direction at any point remain constant over time. There are no sudden changes in speed or direction along the streamlines either, so do not draw sharply angled changes in direction when the lines flow around objects.

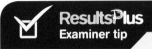

Laminar in a pipe showing streamlines of different but unchanging velocities flow

Turbulent flow is chaotic and subject to sudden changes in speed and direction – eddies are frequently seen. There is a lot of large-scale mixing of layers, so you should include lots of eddies.

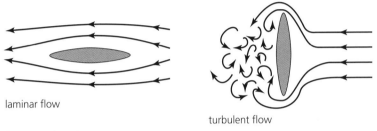

laminar flow

turbulent flow

Flow past obstacles

There may be situations where one type of flow changes to another, such as when a laminar airflow passes around an obstacle and becomes turbulent.

Thinking Task

Why do sports cars have a streamlined shape?

? Quick Questions

Q1 Calculate, in m^3, the volume of:
  a a cube of side 3.4 mm
  b a sphere of radius 17 cm
  c a cylinder of length 26 cm and diameter 56 mm.
Q2 Find the mass of each of the above if:
  a the cube has density 2700 kg m^{-3}
  b the sphere has density 5600 kg m^{-3}
  c the cylinder has density 600 kg m^{-3}.
Q3 A sphere of mass 4.2 kg and volume $5.2 \times 10^{-3}\,m^3$ is placed 1 m below the surface of a liquid of density 1100 kg m^{-3}.
  a Calculate the upthrust on the sphere.
  b Which way will it move?
  c Sketch the shape of the fluid flow just as the sphere starts to move.

Fluids and fluid flow 2

Viscous drag

When solids and fluids move relative to each other, the layer of fluid next to the solid exerts a frictional force on it. Successive layers of fluid experience frictional forces between each other as well. The frictional forces cause **viscous drag**, which is one of the causes of **air resistance**.

Although it acts over much of the surface of a solid object, viscous drag is labelled as a single force in the opposite direction to the relative motion of the fluid past the solid. Viscous drag is greater when the fluid flow is turbulent, so designers of cars, for example, ensure that the flow remains as laminar as possible.

Viscosity

The amount of viscous drag depends on the type of fluid as well as the shape of the object and the type of fluid flow. Viscous drag would be greater in syrup than in water. We say that syrup has a greater **viscosity**, or that it is more **viscous**.

The **coefficient of viscosity**, η, usually just called viscosity, is used to compare different fluids. Fluids with lower viscosity will have a greater rate of flow and cause less viscous drag.

Viscosity is quoted with a variety of units in different contexts. Units include $kg\,m^{-1}\,s^{-1}$, $N\,s\,m^{-2}$ and $Pa\,s$. These are all equivalent, and any would be accepted in an exam answer.

In general, the viscosity of liquids decreases with temperature, and the viscosity of gases increases with temperature. This can be one of the factors used to control the rate of flow, for example, in a factory making chocolates. This also means that, in any experiment you do to measure viscosity, it is vital to record the temperature of the fluid.

Stokes' law

Stokes' law gives a formula for viscous drag for a small sphere at low speeds in laminar flow. The viscous drag force, F is given by:

$$F = 6\pi\eta r v$$

where r is the radius of the sphere, in metres, η is the viscosity, in $N\,s\,m^{-2}$, and v is the velocity, in $m\,s^{-1}$.

Worked Example

A table tennis ball of diameter 40 mm and mass 2.7 g is dragged through water of viscosity $0.001\,N\,s\,m^{-2}$ at a speed of $5\,m\,s^{-1}$. Calculate the viscous drag and comment on your answer.

- -

$$\text{radius} = \frac{40\,mm}{2} = 20\,mm = 0.02\,m$$

so viscous drag is

$$F = 6\pi\eta r v$$
$$= 6 \times \pi \times 0.001\,N\,s\,m^{-2} \times 0.02\,m \times 5\,m\,s^{-1}$$
$$= 1.8 \times 10^{-3}\,N$$

This is a very small force, much less than the weight of the ball, suggesting that the conditions for Stokes' law do not apply.

Terminal velocity

The diagram summarises the forces on a sphere falling through a fluid. The resultant downwards force on the sphere is weight – upthrust – viscous drag. Upthrust and weight are constant and the viscous drag is proportional to the downward velocity of the sphere.

If the weight is greater than the upthrust the sphere will accelerate downwards. As it accelerates, viscous drag increases and the resultant force decreases. The limit to this is when

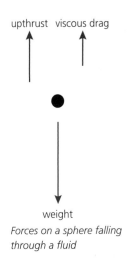

Forces on a sphere falling through a fluid

$$\text{weight} - \text{upthrust} - \text{viscous drag} = 0$$

or

$$\text{upthrust} + \text{viscous drag} = \text{weight}$$

As the resultant force is zero, there is no more acceleration. The velocity at which this occurs is known as the **terminal velocity**.

The terminal velocity of a sphere in a fluid can be measured relatively easily. This is often done by dropping a small steel ball bearing through the fluid. This then enables the viscous drag, and hence the viscosity, to be determined.

The terminal velocity can be expressed in terms of Stokes' law. At the terminal velocity, upthrust + viscous drag = weight. Therefore,

$$\frac{4}{3}\pi r^3 \rho_{\text{fluid}}\, g + 6\pi\eta r v = \frac{4}{3}\pi r^3 \rho_{\text{steel}}\, g$$

Rearranging this further gives

$$v = \frac{2r^2 g(\rho_{\text{steel}} - \rho_{\text{fluid}})}{9\eta}$$

⟨?⟩ Quick Questions

Q1 Explain carefully the effect that an increase in temperature will have on the terminal velocity of a small ball bearing falling through vegetable oil.

Q2 A toy helium balloon of radius 30 cm is released. The weight of the balloon is 0.17 N, the upthrust is 0.18 N and the viscosity of the surrounding air is $1.8 \times 10^{-5}\,\text{N s m}^{-2}$.
  a Sketch the forces acting.
  b Use this to help you find the terminal velocity using Stokes' law.
  c Comment on your answer.

ResultsPlus
Watch out!

Some students get so used to the diagram of an object falling through a fluid that they always draw viscous drag upwards, instead of in the opposite direction to the motion.

Force and extension

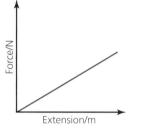

Graph of force against extension for a spring

Hooke's law

The graph shows force against extension for a spring. As the straight-line graph illustrates, force is directly proportional to extension:

$$F = k\Delta x$$

where Δx is the change in length x, or the extension. The constant is k, known as the **spring constant** or **stiffness**, with units $N\,m^{-1}$.

The spring is said to obey **Hooke's law** (that force is directly proportional to extension) within this range of forces. It also behaves elastically, because when the force is removed it returns to its original dimensions.

For many springs, as long as their coils aren't touching at the start, the same applies when they are compressed. The extension is now a decrease in length (sometimes called *compression* rather than *extension*).

Worked Example

a A spring has an extension of 20 cm when the applied force is 14 N. Find the spring constant.

b Calculate the force to achieve an extension of 27 cm. State the assumption you make.

- -

a $k = F/\Delta x = 14\,N/0.2\,m = 70\,N\,m^{-1}$

b $F = k\Delta x = 70\,N\,m^{-1} \times 0.27\,m = 19\,N$

This assumes that extension remains proportional to force for the new extension, i.e. Hooke's law is obeyed.

Describing materials

Stiffness is one way of describing a material. You will need to understand the meaning of these other words:

- **brittle** materials break with little or no plastic deformation
- **malleable** materials can be beaten into sheets; these materials show a large plastic deformation under compression
- **ductile** materials can be pulled into wires or threads; these materials show plastic deformation before failure under tension
- **hard** materials resist plastic deformation by surface indentation or scratching
- **tough** materials can withstand impact forces and absorb a lot of energy before breaking; large forces produce a moderate deformation.

Metals, for example, are malleable and ductile, and glass is brittle and hard.

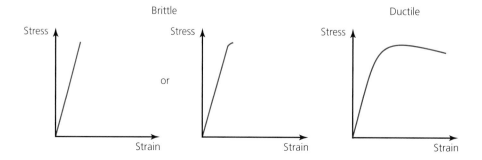

Limits

Only some materials follow Hooke's law, and others only do so up to a certain point. The point beyond which force is no longer proportional to extension is the **limit of proportionality**. After this point, the force-extension graph is no longer straight.

For springs and many materials, there is a short further region where the behaviour is still elastic, which means they return to their original length when the force is removed. This ends at the **elastic limit**, after which further force produces some permanent deformation. This is known as the **plastic** region.

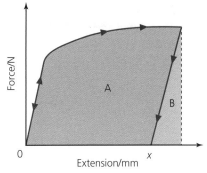

Extending a spring beyond the elastic limit

In the diagram, when the force is removed the spring returns to extension x rather than to zero – there is a permanent extension.

Elastic strain energy

Work is done by the deforming force in extending it, and the energy is stored in the spring as **elastic strain energy**, which is released when the force is removed. The force isn't constant, so work cannot just be found from work = force × displacement.

The shaded area under the upper graph in the diagram is $\frac{1}{2} \times F \times \Delta x$. This is the same as average force × displacement, i.e. work done, so we can say that elastic strain energy stored, E_{el} = area under the graph. This applies to any graph, e.g. the shaded area for the lower graph, and can be estimated by counting squares. When the graph is a straight line, i.e. Hooke's law applies,

$$E_{el} = \frac{1}{2}F\Delta x$$

Worked Example

Calculate the elastic strain energy stored in the spring in parts **a** and **b** of the previous Worked Example.

- -

a $E_{el} = \frac{1}{2}F\Delta x = \frac{1}{2} \times 14\,\text{N} \times 0.2\,\text{m} = 1.4\,\text{J}$

b As $F = k\Delta x$, then

$$E_{el} = \frac{1}{2}k(\Delta x)^2 = \frac{1}{2} \times 70\,\text{N m}^{-1} \times (0.27\,\text{m})^2 = 2.6\,\text{J}$$

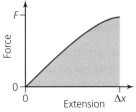

Using the area under a force-extension curve

Quick Questions

Q1 A spring has an extension of 4 cm when a weight of 5 N is suspended from it. What is the extension if three identical springs are suspended side by side and a weight of 10 N is suspended from them?
  A 2.7 cm
  B 5.4 cm
  C 8.0 cm
  D 10.8 cm

Q2 An open-coiled spring extends by 2 cm when a force of 9 N is applied.
  a Calculate the elastic strain energy when it is compressed by 3 cm.
  b The spring has a mass of 18 g. It is allowed to jump into the air freely after being compressed by 3 cm. Calculate the maximum height it could reach.
  c Explain why this is a *maximum* height.

Stress, strain and the Young modulus

The extension of a material depends on the stiffness, k, and the applied force. But it is also the case that for a given force a longer sample will experience a greater extension, and a thicker sample will extend less. If different materials are to be compared fairly, the effect of the sample's thickness and length must be taken into account.

In order to do this, graphs of force against extension are replaced with graphs of **stress** against **strain**.

$$\text{stress} = \frac{\text{applied force}}{\text{cross-sectional area}}$$

$$\sigma = \frac{F}{A}$$

Stress has units Nm^{-2} or Pa (pascal is also the unit of pressure, since pressure is also force/area).

$$\text{strain} = \frac{\text{extension}}{\text{original length}}$$

$$\varepsilon = \frac{\Delta x}{x}$$

Strain has no units, as it is the ratio of two lengths. **Tensile strain** is when a material is being stretched and **compressive strain** is when a material is being compressed.

Worked Example

A wire of diameter 0.86 mm and length 1.5 m has a force of 160 N applied, extending it by 5.2 mm. Calculate **a** tensile stress and **b** tensile strain in the wire.

a Radius of wire is r = 86 mm/2 = 0.86×10^{-3} m/2 = 0.43×10^{-3} m.

$$\text{stress} = \frac{\text{applied force}}{\text{cross-sectional area}} = \frac{\text{force}}{\pi r^2} = \frac{160\,N}{\pi \times (0.43 \times 10^{-3}\,m)^2} = 2.8 \times 10^8\,Pa$$

b Extension of wire is Δx = 5.2 mm = 5.2×10^{-3} m

$$\text{strain} = \frac{\text{extension}}{\text{original length}} = \frac{\Delta x}{x} = \frac{5.2 \times 10^{-3}\,m}{1.5\,m} = 3.5 \times 10^{-3} \text{ (no unit!)}$$

ResultsPlus
Watch out!

Students often get the order of these points mixed up and also their meanings, especially confusing limit of proportionality with elastic limit, and confusing elastic limit with yield point.

Stress/strain graphs

The diagram shows a typical graph for a metal wire.

- P is the **limit of proportionality**. The line is straight up to this point because stress is proportional to strain. After P the line is no longer straight, but the material is still behaving elastically and will return to its original length.
- L is the **elastic limit**. Here the wire stops behaving elastically and starts to behave plastically.
- Y is the **yield point**. From here the material shows an appreciably greater increase in strain for a given increase in stress.
- T indicates the maximum tensile stress, which gives its **ultimate tensile strength**, the maximum stress that can be applied before the sample goes on to break.

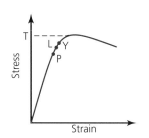

Stress/strain graph for a metal wire

Young modulus

The diagram shows stress-strain graphs for materials A and B. The curve for A is steeper than the curve for B. This shows that material A is stiffer than material B.

The gradient of the stress/strain curve is known as the **Young modulus** of the material.

$$\text{Young modulus} = \frac{\text{stress}}{\text{strain}} = \frac{\text{applied force/cross-sectional area}}{\text{extension/original length}}$$

$$\text{Young modulus} = \frac{\text{applied force} \times \text{original length}}{\text{cross-sectional area} \times \text{extension}}$$

$$E = \frac{Fx}{A\Delta x}$$

The units of the Young modulus are Nm^{-2}, or Pa. For many materials, MPa (10^6 Pa) is a more convenient unit as the Pa is very small.

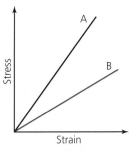

Stress-strain graphs for two different materials

How to measure the Young modulus

The apparatus used to measure the Young modulus is shown in the diagram below. Increase the force applied to a long, thin wire by adding masses of known weight. Measure the diameter of the wire with a micrometer screw gauge, remembering to measure at different positions and in different planes, and find a mean value. Measure the original length, from the clamp to the marker, with a metre rule, and then measure the increase in this length (the extension) using a fixed scale as the force is changed.

Plot a graph of force against extension, and find the gradient, $F/\Delta x$. This is multiplied by the original length/cross-sectional area to find the Young modulus.

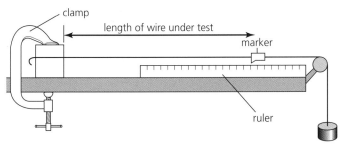

Measuring the Young modulus

Quick Questions

Q1 A wire with Young modulus $3.4 \times 10^{11} \, Nm^{-2}$, diameter 0.84 mm and length 1.7 m has a force of 125 N applied.
  a Calculate the tensile stress.
  b Calculate the extension produced.
Q2 Rearrange the instructions for the Young modulus experiment, adding any extra details of your own, into bullet points covering equipment used, measurements taken, processing results and experimental precautions.
Q3 Explain why stress and the Young modulus have the same units as pressure.

Thinking Task

Explain why a long thin wire is used in the Young modulus experiment.

Materials checklist

By the end of this section you should be able to:

Revision spread	Checkpoints	Spec. point	Revised	Practice exam questions
Fluids and fluid flow 1	Understand and use the terms density, laminar flow, streamline flow, turbulent flow and upthrust, for example, in transport design or in manufacturing	18	☐	☐
	Recognise and use the expression upthrust = weight of fluid displaced	20	☐	☐
Fluids and fluid flow 2	Understand and use the terms terminal velocity and viscous drag, for example, in transport design or in manufacturing	18	☐	☐
	Recall that the rate of flow of a fluid is related to its viscosity	19	☐	☐
	Recognise and use the expression for Stokes' law, $F = 6\pi\eta r v$	20	☐	☐
	Recall that the viscosities of most fluids change with temperature, and explain the importance of this for industrial applications	21	☐	☐
Force and extension	Draw force/extension and force/compression graphs	22	☐	☐
	Use Hooke's law, $F = k\Delta x$, and know that it applies only to some materials	23	☐	☐
	Investigate elastic and plastic deformation of a material, and distinguish between them	25	☐	☐
	Explain what is meant by the terms brittle, ductile, hard, malleable, stiff and tough, and give examples of materials exhibiting such properties	26	☐	☐
	Calculate the elastic strain energy E_{el} in a deformed material sample, using the expression $E_{el} = \frac{1}{2}Fx$, and from the area under its force/extension graph	27	☐	☐
Stress, strain and the Young modulus	Draw tensile (compressive) stress/strain graphs, and identify the limit of proportionality, elastic limit and yield point	22	☐	☐
	Explain the meaning of, use and calculate tensile (compressive) stress, tensile (compressive) strain, strength, breaking stress, stiffness and Young modulus	24	☐	☐
	Obtain the Young modulus for a material	24	☐	☐

ResultsPlus
Build Better Answers

A child's birthday balloon is filled with helium to make it rise. A ribbon is tied to it, holding a small plastic mass designed to prevent the balloon from floating away.

a Add labelled arrows to the diagram of the balloon to show the forces acting on the balloon. **[2]**

✓ Examiner tip	Student answer
Always draw arrows representing forces where they act, if possible. And remember that weight is a force; gravity is what causes weight. Force arrows should never be labelled as 'gravity'.	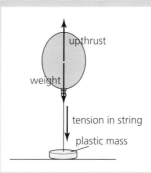

b The balloon is approximately a sphere, of diameter 30 cm. Show that the upthrust on the balloon is about 0.2 N. The density of the surrounding air $\rho = 1.30 \, \text{kg m}^{-3}$. **[3]**

✓ Examiner tip

This question asks you to show how you work out a value. If your answer is nothing like the suggested one, go back and check your working! Always quote more significant figures than the value given to you in the question.

Student answer	Examiner comments
upthrust $= \rho g V = 1.30 \, \text{kg m}^{-3} \times 9.81 \, \text{m s}^{-2} \times \frac{4}{3} \pi (0.15 \, \text{m})^{-3} = 0.18 \, \text{N}$	Don't forget to halve the diameter – many students get the wrong answer because they use the diameter instead of the radius.

c The ribbon is cut and the balloon begins to rise slowly. Sketch a diagram to show the airflow around the balloon as it rises. **[1]**

Student answer	Examiner comments
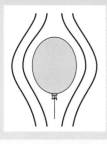	Streamlines representing laminar flow should be smooth, and should not cross each other. Always draw at least three lines.

d A student suggests that, if the balloon reaches terminal velocity, its motion could be described by the relationship

$$mg + 6\pi\eta v = \frac{4}{3}\pi r^3 \rho g$$

where η = viscosity of air, m = mass of the balloon, r = radius of the balloon and v = the terminal velocity reached.
Write the above relationship as a word equation. **[1]**

Student answer	Examiner comments
weight + Stokes' law = volume × density	No marks. There is only one mark for this part of the question, so you have to get all the terms correct to gain the mark. Weight is correct for the first term. The second term is Stokes' law, but it represents *viscous drag*. The third term is the upthrust. So weight + drag = upthrust

Adapted from *Edexcel January 2008 PSA2*

Practice exam questions

1 Skydivers can change the speed at which they fall by changing their body position.

Which statement is *not* true?

A Y will fall faster than X.
B X is more likely to produce turbulent flow than Y.
C Viscous drag acts upwards for X and Y.
D The cross-sectional area is greater for Y than for X.

2 Stokes' law cannot be used to calculate the drag on the skydiver in question 1. Which is the best explanation for this statement?

A The skydiver is too big.
B Stokes' law only applies to laminar flow.
C Stokes' law only applies to turbulent flow.
D The velocity of the skydiver changes during the jump.

3 What is the upthrust on a 1 cm diameter steel ball in water?
Density of steel = $8100 \, \text{kg m}^{-3}$, density of water = $1000 \, \text{kg m}^{-3}$.

A $2.3 \times 10^{-1} \, \text{N}$
B $3.4 \times 10^{-1} \, \text{N}$
C $4.2 \times 10^{-2} \, \text{N}$
D $5.1 \times 10^{-3} \, \text{N}$

4 Volcanoes vary considerably in the strength of their eruptions. A major factor in determining the severity of the eruption is the viscosity of the magma material. Magma with a high viscosity acts as a plug in the volcano, allowing very high pressures to build up. When the volcano finally erupts, it is very explosive. Once magma is out of the volcano, it is called lava.

   a How would the flow of high-viscosity lava differ from that of lava with a low viscosity? **[1]**
   b What would need to be measured to make a simple comparison between the viscosities of two lava flows? **[1]**
   c When the lava is exposed to the atmosphere, it cools rapidly. What effect would you expect this cooling to have on the lava's viscosity? **[1]**
   d When lava is fast flowing, changes to its viscosity disrupt the flow, making it no longer laminar. Use labelled diagrams to show the difference between laminar and turbulent flow. **[3]**
   e Different types of lava have different viscosities. The least viscous type has a viscosity of about $1 \times 10^3 \, \text{N s m}^{-2}$ whereas a silica-rich lava has a viscosity of $1 \times 10^8 \, \text{N s m}^{-2}$. What type of scale would be used to display these values on a graph? **[1]**

Edexcel June 2007 PSA2

5 a A high-strength concrete has been developed that can withstand a maximum compressive stress of 800 MPa, double that of steel. A sample of this concrete has a cross-sectional area of 20 m². Calculate the maximum force that it could be subjected to before breaking. **[2]**

b This concrete has a Young modulus of 5.0×10^9 Pa. Calculate its strain when under maximum compressive stress. **[2]**

c Unfortunately, concrete has a relatively low tensile strength. This could result in the concrete cracking. What word describes the behaviour of the concrete in this case? **[1]**

Edexcel January 2007 PSA2

6 It is common for pens to have retractable ink refills. When a force F is applied to the button at the end of the pen, the tip of the refill is pushed out of the body of the pen. This compresses a spring in the end of the pen so that if the button is pressed again the refill is pushed back inside the pen.

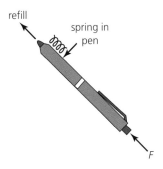

a What sort of deformation must the spring undergo when compressed? Justify your answer. **[2]**

In an experiment, an increasing force was used to compress this spring. The table shows the compression for each value of force.

Force/N	Compression/mm
0.0	0.0
1.0	1.9
2.0	3.8
3.0	5.6
4.0	7.5
5.0	9.4
6.0	11.3
7.0	13.1
8.0	15.0

b Draw a graph to show these results (put force on the x-axis). Add a line of best fit to your points. **[3]**

c Calculate the stiffness of this spring. **[2]**

d In the pen, the spring is compressed by 6.0 mm. What force is needed for this compression? **[1]**

e Calculate the elastic energy stored in the spring when its compression is 6.0 mm. **[3]**

f The spring is replaced by another with double the length but identical in all other ways. How would the force needed to compress this new spring by 6.0 mm compare with the force needed for the original spring? **[1]**

Adapted from Edexcel January 2008 PSA2

Unit 1: Practice unit test

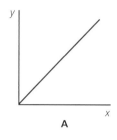

A

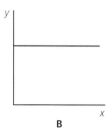

B

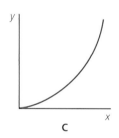

C

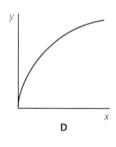

D

Section A

1 A skydiver jumps out of a balloon, which is ascending at $1\,\mathrm{m\,s^{-1}}$. Her total mass, including the parachute, is 70 kg. How far will she have fallen after 5 s (ignoring air resistance)?

    A   118 m
    B   123 m
    C   128 m
    D   133 m   **[1]**

Use the graphs on the left to answer questions 2 and 3.

2 Which of the graphs best represents the quantities described when they are plotted on the y– and x–axes?

     Variable on y-axis: E_{p} for a person in a lift

     Variable on x-axis: Displacement of lift   **[1]**

3 Which of the graphs best represents the quantities described when they are plotted on the y– and x– axes?

     Variable on y-axis: The ratio $\dfrac{F}{\Delta x}$ for a spring

     Variable on x-axis: Load on the spring   **[1]**

4 A trolley of mass 10 kg is rolling down a slope, which is 20° to the horizontal. What is the component of the trolley's weight acting along the slope?

    A   3.4 N
    B   9.4 N
    C   34 N
    D   92 N   **[1]**

5 There are various 'limits' to Hooke's law, called the yield point, the limit of proportionality and the elastic limit. Which statement is *not* correct?

    A   After the yield point is passed, the material shows a greater increase in strain for a given increase in stress.
    B   A stress-strain graph is linear up to the elastic limit.
    C   The material behaves plastically after the elastic limit.
    D   Hooke's law applies only up to the limit of proportionality.   **[1]**

6 A 2 m long wire stretches by 4 mm when a force of 100 N is applied. The diameter of the wire is 0.8 mm. What is the stress on the wire?

    A $0.002\,\mathrm{N\,m^{-2}}$
    B $50\,\mathrm{N\,m^{-2}}$
    C $5 \times 10^{7}\,\mathrm{N\,m^{-2}}$
    D $2 \times 10^{8}\,\mathrm{N\,m^{-2}}$ **[1]**

                                    [6 marks]

Section B

7 The diagrams show a box resting on the floor of a stationary lift and a free-body force diagram showing the forces A and B that act on the box.

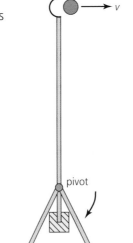

For each of the following situations, say whether forces A and B have increased, decreased or are the same as when the lift is stationary.

  a The lift is accelerating upwards.
  b The lift is moving upwards at a constant speed.
  c The lift is accelerating downwards.
  d The lift is moving downwards at a constant speed.   **[4]**

Adapted from Edexcel January 2007 Unit Test 1

8 A medieval siege engine, called a *trebuchet*, uses a pivoted lever arm to fire a rock projectile. The figure on the left shows a trebuchet which is ready to fire. The gravitational potential energy (E_{grav}) of the large stone counterweight is converted into E_{grav} and kinetic energy (E_k) of the small projectile and E_k of the counterweight.

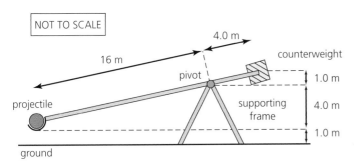

  a The mass of the counterweight is 760 kg. It falls through 5 m. Show that the E_{grav} it loses is about 37 000 J.   **[2]**
  b **i** The mass of the projectile is 55 kg. Its height increases by 20 m as the lever arm rotates. Show that the total E_k of the projectile and the counterweight is about 26 000 J.   **[2]**
      ii State one assumption you have made in your calculation of the E_k. **[1]**
     iii The equation below can be used to find the speed v.

$$26\,000\,\text{J} = \frac{1}{2} \times 760\,\text{kg} \times \left(\frac{v}{4}\right)^2 + \frac{1}{2} \times 55\,\text{kg} \times v^2$$

        Explain the term $\frac{1}{2} \times 760\,\text{kg} \times \left(\frac{v}{4}\right)^2$ in this equation.   **[2]**

  c Solving this equation gives a speed v of 22.5 m s⁻¹.
    i Assuming the trebuchet launches its projectile horizontally over level ground, calculate the time of flight of the projectile.   **[2]**
    ii Calculate the distance the projectile travels before it hits the ground.   **[2]**

Edexcel January 2008 PSA1

9 A cyclist and a car are both stationary at traffic lights. They are alongside each other with their front wheels in line. The lights change and they both move forward in the same direction along a straight flat road. The idealised graph shows the variation of velocity against time for both the cyclist and the car from the instant the lights change to green to the instant they are again level.

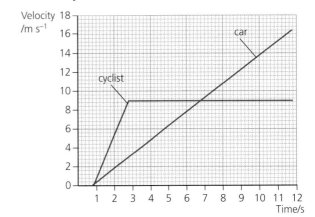

  a What does the time interval of 0.8 s at the beginning of the graph represent?   **[1]**

b i How long does it take, from the instant the lights change to green, for the car to reach the same velocity as the cyclist? **[1]**
ii Determine the distance between the cyclist and the car at this time. **[3]**
c What is the relationship between the average velocity of the cyclist and the average velocity of the car for the time interval covered by the graph? **[1]**

Edexcel January 2007 Unit Test 1

10 A cricketer bowls a ball from a height of 2.3 m. The ball leaves the hand horizontally with a velocity u. After bouncing once, it passes just over the stumps at the top of its bounce. The stumps are 0.71 m high and are situated 20 m from where the bowler releases the ball.

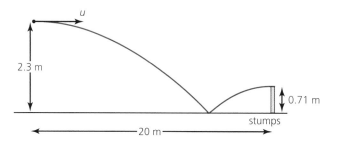

a Show that from the moment it is released, the ball takes about 0.7 s to fall 2.3 m. **[2]**
b How long does it take the ball to rise 0.71 m after bouncing? **[3]**
c Use your answers to parts **a** and **b** to calculate the initial horizontal velocity u of the ball. You may assume that the horizontal velocity has remained constant. **[2]**
d In reality the horizontal velocity would not be constant. State one reason why. **[1]**

Edexcel June 2006 Unit Test 1

11 When going downhill, ski jumpers reach speeds of up to 30 m s^{-1} in order to jump great distances. As they move through the air, their body and ski position determines how far they jump.

a i Use one word to describe the type of airflow that the ski jumper is trying to achieve in mid-air. **[1]**
ii The diagram shows a ski jumper in mid-air. Make a rough copy of the diagram and sketch the airflow pattern. **[2]**
iii Suggest one way in which the ski jumper's equipment is designed to produce the maximum possible speed. **[1]**
b Below is a list of material properties. Select one that is desirable and one that is undesirable for material from which the jumper's skis are made. Explain your choices. **[4]**

elastic   tough   plastic

Edexcel January 2007 PSA2

12 Kevlar is a very useful and *strong* material. It can withstand temperatures of up to 300°C and shows no loss of strength or signs of becoming *brittle* at temperatures as low as −196°C. It undergoes *plastic deformation* when subjected to a sudden force and is used to make a variety of objects, from bullet-proof vests to bicycle tyres and canoes.

a Explain what is meant by the words in *italic* in the above passage. **[3]**

One type of Kevlar, Kevlar 49, is used to make cloth. The table gives details of some properties of a single fibre of Kevlar 49.

Diameter/mm	Breaking stress/10^9 Pa
0.254	3.80

b Use information from the table to show that the maximum force which can be exerted on a single fibre of Kevlar 49 without breaking it is about 200 N.   **[3]**

c The material has a Young modulus of 1.31×10^{11} Pa. Calculate the extension of the Kevlar 49 fibre when a stress of 2.00×10^9 Pa is applied to a 1.10 m length.   **[3]**

Edexcel June 2007 PSA2

13 A child's trampoline consists of a plastic sheet tied to a frame by an elastic rope as shown below. When a child jumps on the trampoline, the rope stretches allowing the plastic sheet to move up and down.

Before the trampoline was constructed the rope was tested by hanging weights from one end and measuring the extension. The following table shows the results obtained.

Weight/N	Extension/m
20	0.80
40	1.79
60	2.72
80	3.63
100	4.55
120	5.46
140	6.48

a i Plot a graph to show the results. Put extension on the x-axis.   **[2]**

  ii Draw a line of best fit on the graph and then use the graph to calculate the stiffness of the rope.   **[3]**

b Calculate the energy stored in the rope when a weight of 150 N is hung on the end.   **[2]**

c In using the trampoline, a child weighing 150 N does not cause the rope to extend as much as shown on the graph. Suggest a reason for this.   **[1]**

d This trampoline could be adapted so that an adult could use it by replacing the elastic rope with one made of the same material but with different dimensions. Explain how the new dimensions would have to be different from those of the original rope.   **[2]**

Adapted from *Edexcel June 2008 PSA2*

[56 marks]
[Total 62 marks]

Waves

A travelling wave transfers energy by means of oscillations. Travelling waves can be transverse (when the oscillations are at right angles to the direction of wave travel) or longitudinal (when the oscillations are parallel to the direction of travel).

- The **amplitude** of a wave is the maximum displacement from the equilibrium position.
- The **wavelength** λ of a wave is the shortest distance between two adjacent points that are in phase.
- The **frequency** f of a wave is the number of oscillations per second or the number of waves passing a point in one second.
- The **period** T of a wave is the time taken for one complete oscillation, $T = 1/f$.
- The **speed** v of a wave is given by $v = f\lambda$.

Worked Example

Calculate the frequency of ultraviolet light with wavelength 350 nm. The speed of light in air is $3.0 \times 10^8\,\mathrm{m\,s^{-1}}$.

Rearrange the equation $v = f\lambda$ to give

$$f = \frac{v}{\lambda} = \frac{3.0 \times 10^8\,\mathrm{m\,s^{-1}}}{350 \times 10^{-9}\,\mathrm{m}} = 8.6 \times 10^{14}\,\mathrm{Hz}$$

ResultsPlus
Watch out!

Convert nm to m using
$1\,\mathrm{nm} = 1 \times 10^{-9}\,\mathrm{m}$.

Using graphs to represent waves

A wave can be represented by a graph of displacement against distance along the wave at a particular moment in time. A graph of displacement against time shows the oscillations at one particular position.

Worked Example

Waves are sent along a spring with a velocity of $0.25\,\mathrm{m\,s^{-1}}$. The diagram shows how the displacement varies with distance along the spring at one moment in time. The wave is travelling to the right.

a State the wavelength and calculate the period of the oscillations.

b Sketch graphs of displacement against time for points X and Y.

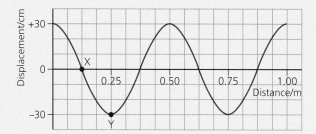

a From the graph the wavelength is 0.50 m.

Calculate the frequency:

$$f = \frac{v}{\lambda} = \frac{0.25\,\mathrm{m\,s^{-1}}}{0.5\,\mathrm{m}} = 0.5\,\mathrm{Hz}$$

Calculate the period:

$$T = \frac{1}{f} = \frac{1}{0.5\,\text{Hz}} = 2.0\,\text{s}$$

b The graph for X starts at a displacement of 0 cm and then initially becomes positive. The graph for Y starts with a displacement of −30 cm. X and Y are a quarter of an oscillation apart, so the time difference between them reaching the same displacements is 0.5 s.

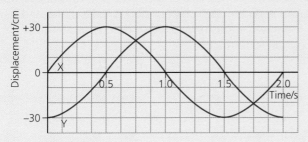

Sound waves

Sound waves are longitudinal waves. Particles of air are displaced from their equilibrium positions and produce regions of compression and rarefaction.

A **wavefront** is a line representing a series of equivalent points on the waves (e.g. all the compressions).

⟨?⟩ Quick Questions

Q1 A sound has a frequency of 5 kHz. The speed of sound in air is 330 m s⁻¹. Calculate **a** the period of the oscillations and **b** the wavelength of these sound waves in air.

Q2 A transverse wave of amplitude 4 cm and frequency 2 Hz travels along a spring with a speed of 1.6 m s⁻¹. Sketch:
  a a displacement-distance graph for a 2.4 m length of the spring (assume the starting point of this length has zero displacement)
  b a displacement-time graph for one point on the spring (assume it starts at maximum positive displacement showing two complete oscillations).

Q3 The diagram shows a wave on a rope. The wave is travelling from left to right. At the instant shown, point L is at a maximum displacement and point M has zero displacement. Describe the motion of points L and M during the next half cycle of the wave.

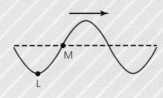

Superposition and standing waves

Two waves are said to be **in phase** if their crests (or troughs) occur in the same place. If the crest of one wave coincides with the trough of another, the waves are **in antiphase**. The phase difference between two waves or two points on a wave is often described using angles. A complete wave is equivalent to an angle of 360° or 2π radians. If waves are in antiphase, they have a phase difference of 180° or π radians.

When two or more waves arrive at the same place at the same time, **superposition** takes place. If the two waves are in phase, the superposition is constructive and produces a large amplitude. If the two waves have a phase difference of 180° (antiphase), then the superposition is destructive.

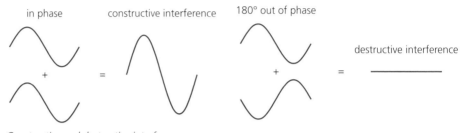

Constructive and destructive interference

Interference

Two sets of waves with the same frequency and a constant phase difference are said to be **coherent** and can produce interference patterns. Interference patterns consist of alternate maxima (where constructive interference is taking place) and minima (where destructive interference is taking place). If both waves are from the same source, the difference in their path lengths determines their phase difference when they arrive at the screen. If the path difference is a whole number of wavelengths, then constructive interference takes place. If the path difference is an odd number of half wavelengths, destructive interference occurs.

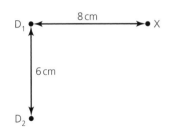

Worked Example

In a ripple tank, two dippers D_1 and D_2, 6 cm apart, produce coherent waves of wavelength 4 cm. Is point X a maximum or minimum?

Find the distance from both sources to X: using Pythagoras' theorem, $(D_2X)^2 = 6^2 + 8^2 = 100$ and so $D_2X = 10$ cm.

Find the path difference from each source to X: path difference = 10 − 8 = 2 cm.

Find the phase difference by comparing the path difference with the wavelength: path difference/wavelength = 2 cm/4 cm = ½ wavelength, so the oscillations are 180° out of phase, giving destructive interference, resulting in a minimum.

ResultsPlus
Examiner tip

In superposition questions, waves must be 'coherent' or 'in phase' to give an observable interference pattern.

Standing waves

Superposition of a continuous wave reflected from a boundary with its incident wave will produce an interference pattern called a **standing wave** or **stationary wave**. The maxima are called **antinodes** and the minima are called **nodes**. The distance between adjacent minima is half the wavelength of the wave.

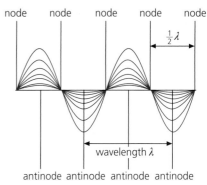

Standing waves

Standing waves on strings and in pipes

A standing wave on a string has a node at each end. For sound waves travelling in a tube, a closed end produces a displacement node and an open end produces a displacement antinode.

Worked Example

An organ pipe, open at both ends, is 2.0 m long. Calculate the frequency of the fundamental note produced. The speed of sound in air is 330 m s^{-1}.

Calculate the wavelength: $\lambda = 2 \times 2.0\,\text{m} = 4.0\,\text{m}$.

Calculate the frequency:

$$f = \frac{v}{\lambda} = \frac{330\,\text{m s}^{-1}}{4.0\,\text{m}} = 82.5\,\text{Hz}$$

❓ Quick Questions

Q1 The wind is blowing across the end of a pipe 1.5 m long and sets up a standing wave. The pipe is closed at one end. What is the lowest frequency of sound that will be produced?

Q2 Light from a laser is shone through two slits onto a screen. The screen shows alternate blobs of light and dark. Explain this pattern.

Q3 A standing wave is produced in a pipe that is open at both ends. Describe the standing wave in this pipe in terms of the numbers of nodes and antinodes and their positions.

⚙ Thinking Task

Light reflects from the top and bottom of a layer of oil on a puddle.

a Explain why the path difference between the reflected rays is twice the thickness of the oil layer.

b The path difference at some places on the oil is exactly equal to the wavelength of red light. Explain what will be seen at these positions.

c Why will different colours of light be seen at different places on the puddle?

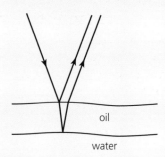

The electromagnetic spectrum

Visible light is just one type of wave in the electromagnetic spectrum. All the waves have a number of properties in common: they all travel at the same speed in a vacuum (3×10^8 m s^{-1}), they are all transverse waves, and they all consist of an oscillating electric and magnetic field.

ResultsPlus
Examiner tip

Ensure that you know the sequence of the types of wave in the spectrum. Gamma rays have the shortest wavelength and largest frequency.

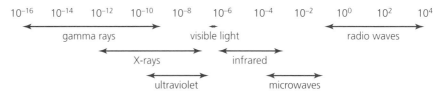

| 10^{-16} | 10^{-14} | 10^{-12} | 10^{-10} | 10^{-8} | 10^{-6} | 10^{-4} | 10^{-2} | 10^{0} | 10^{2} | 10^{4} |

gamma rays   visible light   radio waves

X-rays   infrared

ultraviolet   microwaves

Types of electromagnetic radiation, showing approximate wavelengths in metres

Worked Example

Calculate the wavelength of an electromagnetic wave of frequency 2.0×10^{19} Hz. Hence establish which type of electromagnetic wave this is.

- -

Rearrange the equation $v = f\lambda$ to give

$$\lambda = \frac{v}{f} = \frac{3.0 \times 10^8 \text{ m s}^{-1}}{2.0 \times 10^{19} \text{ Hz}} = 1.5 \times 10^{-11} \text{ m}$$

Using the spectrum above, these waves are X-rays.

Applications of electromagnetic waves

These are just some of the applications of different parts of the electromagnetic spectrum.

Radio waves are used for TV and radio transmissions. The waves used for TV have wavelengths of about 1 m. This means that TV signals cannot diffract past a large object such as a hill, so a house in a valley would not receive the signal from a transmitter on the other side of a hill. Another transmitter on top of the hill would be required – this is called line-of-sight transmission.

Microwaves are used for mobile phone links and communications via satellite. The frequency of the transmitted wave from a mobile phone is different from the frequency of the received wave. This is because the two waves could superpose and produce a standing interference pattern if the frequencies were the same.

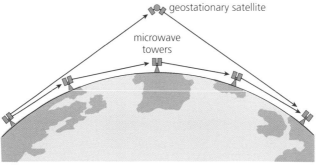

Microwaves are transmitted along 'lines of sight' or via satellites. Their short wave length means they cannot diffract round obstacles.

Infrared waves are used by some alarm systems. Warm bodies emit infrared waves. These can be detected by sensors, which can then activate an alarm system.

Visible light is used by humans and most animals to receive information about their surroundings.

Ultraviolet (UV) waves can cause fluorescence. Some materials will emit visible light when irradiated by ultraviolet waves. Some night-clubs stamp a dye onto your wrist that is invisible under normal light but emits blue light when placed under a UV light. If you go out and want to re-enter the club, a UV lamp is used to check that you have the dye on your wrist. UV waves detected from stars and galaxies are also used to study these objects.

X-rays are penetrating. They can be used to image objects inside a suitcase at an airport or to image the inside of the body.

Gamma rays are used in hospitals to sterilise equipment. Gamma 'bursts' can also be detected from some distant galaxies and are thought to be produced by massive supernova explosions. If they were produced in our galaxy, they would cause significant climate change and disrupt evolution on Earth.

(?) Quick Questions

Q1 BBC Radio 4 uses radio waves with a frequency of 198 kHz. Calculate their wavelength and suggest why only one transmitter is needed for the whole of England.

Q2 Why do you think UV lamps are sometimes called 'black lights'?

Q3 List some ways in which different parts of the electromagnetic spectrum are used to detect things that we cannot see directly.

Thinking Task

Explain how microwaves of the same frequency sent and received by a mobile phone could lead to interference.

Pulse–echo techniques

Waves reflect from a boundary between two media. The greater the difference in density between the two materials, the stronger the reflection. Pulse–echo techniques are used to detect the position and/or the motion of a boundary between two materials. A pulse is required so that the time interval between the incident pulse and the reflected pulse (echo) can be measured.

Finding position

Radars use radio waves to detect the positions of aeroplanes.

ResultsPlus
Watch out!

Don't forget to halve the distance the waves have travelled, as the waves have travelled there and back.

> **Worked Example**
>
> A radar sends a pulse of radio waves, which returns 0.20 ms later. Calculate the distance to the reflecting object. The speed of radio waves is $3 \times 10^8 \, \text{m s}^{-1}$.
>
> We know that
>
> $$\text{distance} = \text{speed} \times \text{time} = 3 \times 10^8 \, \text{m s}^{-1} \times 0.20 \, \text{ms} = 60\,000 \, \text{m}$$
>
> So distance to the reflecting object is 30 km.

Finding the speed of a moving boundary

When waves are emitted from a moving source or detected by a moving receiver, the detected frequency differs from the emitted frequency. This is called the **Doppler effect**. The shift in frequency is proportional to the relative speed of the motion.

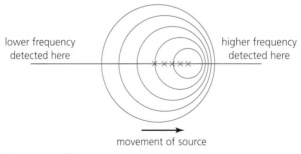

lower frequency detected here

higher frequency detected here

movement of source

The Doppler effect

> **Worked Example**
>
> A radar trap emits a pulse of radio waves of frequency 24 GHz. A car moving at 30 mph causes a change of frequency of 1070 Hz. The reflected pulse from a car in a 30 mph zone has a wavelength of $1.249\,9999 \times 10^{-2}$ m. Was the car moving towards or away from the radar trap, and was it speeding?
>
> Calculate the received frequency using
>
> $$f = \frac{v}{\lambda} = \frac{3 \times 10^8 \, \text{m s}^{-1}}{1.249\,9999 \times 10^{-2} \, \text{m}} = 2.400\,0002 \times 10^{10} \, \text{Hz} = 24.000\,002 \, \text{GHz}$$
>
> As the frequency has increased from 24 GHz, the car was moving towards the source.
>
> The change in frequency is 2000 Hz. This is larger than the change at 30 mph, so this car was speeding at approximately double the speed limit.

Ultrasound scanning

An ultrasound scan of a foetus is usually taken at about 12 weeks. Reflected pulses of ultrasound are used to determine where the boundaries are between different tissues and then build an image.

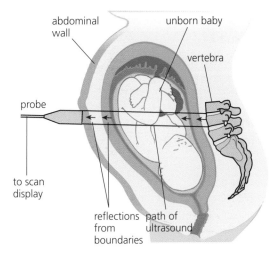

Ultrasound scanning

The **resolution** (the smallest level of detail that can be seen) can be improved by reducing the wavelength of the sound used. Resolution can also be improved by using pulses of very short time interval.

Techniques such as ultrasound scanning must be tested carefully before they are used, to ensure that they do not harm the mother or baby. X-rays used in medicine can harm the patient, so doctors must decide whether the benefits from using X-rays to diagnose a problem outweigh the possible harm the X-rays may cause.

(?) Quick Questions

Q1 In 1990 the surface of Venus was imaged using radio waves sent by an orbiting satellite. The images had a resolution of 100 m of vertical height.
 a Explain how waves can be used to determine the height of the surface of the planet.
 b What factors would improve the resolution?

Q2 How do radar speed guns detect the speed of oncoming cars?

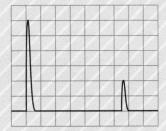

Q3 The result of an ultrasound scan of a human organ is shown. The speed of sound in the organ is 1400 m s^{-1}. The smallest grid division on the x-axis is 1×10^{-5} s.
 a Calculate the thickness of the organ.
 b Why is a gel of similar density to skin smeared on the ultrasound emitter as it is placed on the skin of the person?

Refraction

When waves meet a boundary between two materials, some of the wave is reflected and some is transmitted. The transmitted wave changes speed and may change direction. This is **refraction**.

The refractive index between two materials 1 and 2 is:

$$_1\mu_2 = \frac{\text{speed of wave in material 1}}{\text{speed of wave in material 2}} = \frac{v_1}{v_2}$$

or

$$_1\mu_2 = \frac{\text{sine of angle in material 1}}{\text{sine of angle in material 2}} = \frac{\sin i}{\sin r}$$

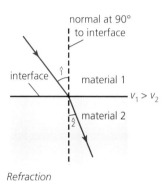

Refraction

Worked Example

A student uses a ray box and a glass prism to study refraction. She measures an angle in air of 35° and a corresponding angle in the glass of 23°. Calculate the refractive index of the glass.

$$_{air}\mu_{glass} = \frac{\text{sine of angle in air}}{\text{sine of angle in glass}} = \frac{\sin 35°}{\sin 23°} = 1.47$$

Total internal reflection

If a wave passes from a more dense to a less dense material, it is possible for all the light to reflect and none to refract at that interface. This is called **total internal reflection**. This happens if the angle of incidence within the material is greater than the **critical angle** C. At smaller angles, some of the light may be reflected.

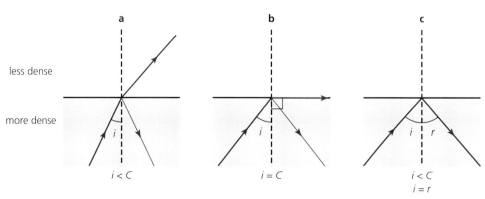

Refraction and total internal reflection

The critical angle is related to the refractive index. Note that, when the angle in the material $= C$, then $_1\mu_2 = \sin 90°/\sin C$, and as $\sin 90° = 1$ this simplifies to:

$$_1\mu_2 = \frac{1}{\sin C}$$

Worked Example

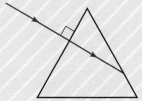

The refractive index for light travelling from water to glass is 1.15. Calculate the critical angle and state the material in which total internal reflection will occur.

Using $_1\mu_2 = 1/\sin C$ gives $\sin C = 1/1.15$, and so $C = 60.4°$. Total internal reflection can only happen when light increases speed. This happens in glass.

(?) Quick Questions

Q1 A ray of light is incident at 90° on one face of an equilateral prism as shown. The critical angle for glass–air is 42°. Copy the diagram and complete it to show the path of the ray.

Q2 The refractive index for light travelling from air to Perspex is 1.4. The speed of light in air is $3 \times 10^8\,\mathrm{m\,s^{-1}}$.
   a Calculate the speed of light in Perspex.
   b A ray of light in air has an incident angle of 40°. Calculate the refracted angle in Perspex.
   c Draw a diagram to illustrate this ray of light passing from air to Perspex.

⚙ Thinking Task

The diagram shows a ray of light entering a spherical rain drop centre O.
   a Take suitable measurements to show that the refractive index of water is about 1.3.
   b Calculate the critical angle for water.
   c Use another measurement taken from the diagram to explain whether the reflection at X is partial or total.
   d A rainbow consists of a spectrum of colours. What does this suggest about the refractive index of water?

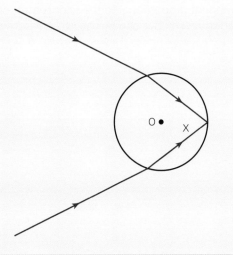

Polarisation

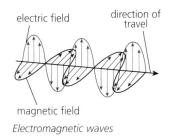

Electromagnetic waves

Light is a transverse wave. It consists of varying electric and magnetic fields at right angles to its direction of motion.

- In unpolarised light, these variations take place in all planes at right angles to the direction in which the ray of light is travelling.
- In plane polarised light, the variations in electric field take place only in one plane. The variations in magnetic field are in a plane at right angles to this.

Longitudinal waves such as sound cannot be polarised.

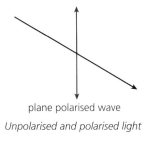

Unpolarised and polarised light

Polarising filters

If unpolarised light encounters a polarising filter, some of it is absorbed and the emerging light is polarised.

If polarised light encounters a polarising filter, polarised light emerges, and its brightness and plane of polarisation depend on the orientation of the filter.

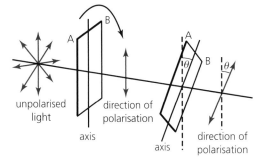

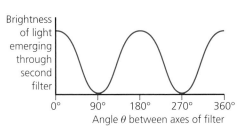

The effects of polarising filters

If two polarising filters are arranged so that they are orientated at right angles to each other, then they will completely absorb unpolarised light. The filters are said to be crossed.

Optical activity

Optically active substances such as sugar solutions rotate the plane of polarisation by an amount proportional to their concentration and the depth of liquid through which the light travels. This can be used to measure the concentration of sugar solutions.

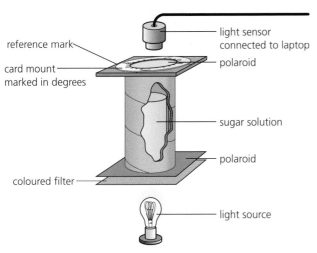

reference mark
card mount marked in degrees
coloured filter
light sensor connected to laptop
polaroid
sugar solution
polaroid
light source

Apparatus used to measure the rotation of the plane of polarisation

 **Quick Questions**

Q1 A student holds up a polarising filter and looks through it at a clear blue sky. She rotates the filter and notices that the sky appears significantly darker with certain positions of the filter. Explain the conclusions you can draw about the light from the blue sky.

Q2 Polarising sunglasses can significantly reduce the glare of sunlight reflected off a road surface. Explain how the polarising filters in such sunglasses can reduce this reflected light.

Q3 A plane polarised light wave is viewed through a polarising filter. Light can be seen for the current position of the filter. The filter is then rotated. Which one of the following angles would the filter be rotated through so that the light is seen again?

A 90°   **B** 270°   **C** 450°   **D** 540°

Thinking Task

The light passing through a laptop screen is polarised. The screen produces images when segments of the screen are switched to become polarising filters at right angles to the plane of the polarised light.

a Explain the difference between unpolarised and polarised light.

b Explain how the segments of the screen produce an image.

c A laptop screen can have its image made secret by placing a large polarising filter on top of it. Explain how this works.

Diffraction

Diffraction is the spreading of wavefronts as a wave passes through a gap or around an obstacle. The amount of diffraction is greatest when the wavelength is similar to the size of the gap or obstacle.

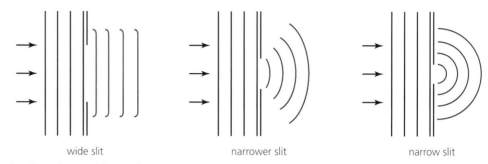

| wide slit | narrower slit | narrow slit |

The effect of gap width on diffraction

Worked Example

The entrance from the sea to many harbour towns is often through a gap between two solid jetties. Explain what happens to violent waves in a storm as they pass through the gap and why the fishing fleet will be protected.

Most diffraction will occur when the gap is approximately equal to the wavelength of the waves. The water waves will spread out as they enter the harbour. Their energy will be spread out over a larger arc length, so the height of the waves will decrease. By the time they reach the fishing boats, the height of the waves will have decreased and they will cause less damage.

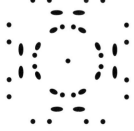

An X-ray diffraction pattern

X-ray and electron diffraction

X-rays can be used to explore the arrangement of atoms within solid materials. The layers of atoms within solid materials can act as a series of gaps or slits. The X-rays first diffract as they pass through the gap. Superposition of X-rays from different layers will then produce an interference pattern. If there is a regular pattern, then the atoms in the solid are regularly structured (ordered).

Electrons can also be used to investigate materials. When a beam of electrons is directed at a crystal, a pattern like that shown below can be observed.

The electron beam must have diffracted through the layers of atoms in the crystal, which suggests that electrons can behave like waves. The wavelength of the electron beam must be similar to the distance between the layers of atoms in the crystal. Electrons were initially thought to be particles, but diffraction experiments confirmed that they can also behave as waves.

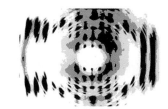

An electron diffraction pattern made by a crystal

(?) Quick Questions

Q1 Explain why visible light (wavelength range 4×10^{-7} m to 7×10^{-7} m) passes straight through a quartz crystal whereas X-rays passing through the same crystal produce a diffraction pattern.

Q2 An electron beam is directed at an unstretched elastic band. The electrons produce no discernible pattern. The band is then stretched and a set of rings appears. Explain what you can deduce about the arrangement of the molecules in an unstretched band compared with the stretched band.

Q3 In an experiment to demonstrate interference using red light, two very narrow slits are arranged so that they are very close together. A series of dots of light are seen on the screen.

  a Explain why the slits need to be very narrow.

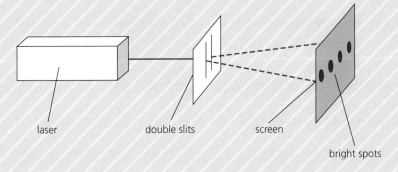

laser   double slits   screen

bright spots

The two slits are then gradually moved further apart. Initially the pattern of dots on the screen move closer together, but eventually the pattern disappears.

  b Explain why the dots eventually disappear.

⚙ Thinking Task

A student argues that electrons are particles because we can measure the mass of an electron. Another student argues that electrons are waves. Describe the evidence for electrons behaving as waves and suggest whether either student is right or wrong.

Waves checklist

By the end of this section you should be able to:

Revision spread	Checkpoints	Spec. point	Revised	Practice exam questions
Waves	Explain and use the concept of wavefronts	33	☐	☐
	Understand and use the terms amplitude, frequency, period, speed and wavelength	28	☐	☐
	Use the wave equation $v = f\lambda$	30	☐	☐
	Recall that a sound wave is a longitudinal wave which can be described in terms of the displacement of molecules	31	☐	☐
	Use graphs to represent transverse and longitudinal waves	32	☐	☐
Superposition and standing waves	Explain and use the concepts of coherence, path difference, superposition and phase	33	☐	☐
	Recognise and use the relationship between phase difference and path difference	34	☐	☐
	Explain what is meant by a standing (stationary) wave, investigate how such a wave is formed, and identify nodes and antinodes	35	☐	☐
	Use graphs to represent standing waves	32	☐	☐
The electromagnetic spectrum	Identify the different regions of the electromagnetic spectrum and describe some of their applications	29	☐	☐
	Explore how science is used by society to make decisions	72	☐	☐
Pulse–echo techniques	Explore and explain how a pulse–echo technique can provide details of the position and/or speed of an object	46	☐	☐
	Describe applications that use a pulse–echo technique	46	☐	☐
	Explain qualitatively how the movement of a source of sound or light relative to an observer/detector gives rise to a shift in frequency (Doppler effect) and explore applications that use this effect	47	☐	☐
	Explain how the amount of detail in a scan may be limited by the wavelength of the radiation or by the duration of the pulses	48	☐	☐
	Discuss the social and ethical issues that need to be considered, e.g. when developing and trialling new medical techniques on patients or when funding a space mission	49	☐	☐
Refraction	Recall that, in general, waves are transmitted and reflected at an interface between media	44	☐	☐
	Explain how different media affect the transmission/reflection of waves travelling from one medium to another	45	☐	☐
	Recognise and use the expression for refractive index $_1\mu_2 = \sin i / \sin r = v_1/v_2$, determine refractive index for a material in the laboratory, and predict whether total internal reflection will occur at an interface using critical angle	36	☐	☐
	Investigate and explain how to measure refractive index	37	☐	☐
	Discuss situations that require the accurate determination of refractive index	38	☐	☐
Polarisation	Investigate and explain what is meant by plane polarised light	39	☐	☐
	Investigate and explain how to measure the rotation of the plane of polarisation	40	☐	☐
Diffraction	Investigate and recall that waves can be diffracted and that substantial diffraction occurs when the size of the gap or obstacle is similar to the wavelength of the wave	41	☐	☐
	Explain how diffraction experiments provide evidence for the wave nature of electrons	42	☐	☐
	Discuss how scientific ideas may change over time, for example, our ideas on the particle/wave nature of electrons	43	☐	☐

ResultsPlus
Build Better Answers

1. A student looks at the sunlight reflected off a puddle of water. She puts a polarising filter in front of her eye. As she rotates the filter, the puddle appears darker then lighter. Explain this observation. **[3]**

☑ Examiner tip

The 3 marks means there are three good points to make. But to gain the 3 marks, you must explain the observation by considering the situation described in the question, i.e. the light from the puddle.

Student answer	Examiner comments
■ The polarising filter cuts out the oscillations if at right angles to them.	■ This **basic answer** made no attempt to explain the observation. Instead, it made a reasonable but general comment about the effect of a filter on polarised light.
▲ The reflected light from the puddle is polarised. Polarised light oscillates in one plane only. If the polarising filter is parallel to this plane, it lets the light through. If it is perpendicular, it blocks the light.	▲ This **excellent answer** does explain the observation in terms of the light from the puddle, and so would get all 3 marks.

2 a In noise cancellation headsets, noise is detected by a microphone in the headset. An electronic circuit 'inverts' this wave and sends this signal to loudspeakers in the headset. Explain how this cancels the noise. **[3]**

☑ Examiner tip

Use the context of the question to add to your answer.

Student answer	Examiner comments
■ Waves that are 180° out of phase with each other cancel.	■ This **basic answer** did not use the context of the question.
▲ Waves from the loudspeaker must have the same frequency and the same amplitude as the noise and be 180° out of phase with each other to cancel by destructive superposition.	▲ This **excellent answer** does specify the origin of the two waves and adds more detail, and so would get all 3 marks.

b In practice, the noise is reduced in intensity rather than cancelled completely. These headsets work well with the noise from a jet engine but are less effective at cancelling speech or music. Explain why. **[2]**

☑ Examiner tip

The first part of a question will often clue you in to the next part. This is true of calculations as well as descriptive answers.

Student answer	Examiner comments
■ Noise from a jet engine is a constant sound so it works better.	■ This **basic answer** was not specific enough.
▲ Noise from a jet engine is a constant frequency whereas speech varies in both frequency and amplitude so by the time a sound wave is produced by the loudspeaker the noise has changed.	▲ This **excellent answer** uses appropriate terminology such as frequency rather than sound and adds more detail, and so would get the 2 marks.

Practice exam questions

1 The diagram below shows a transverse wave. The *y*-axis is displacement in mm, and the *x*-axis is time in ms.

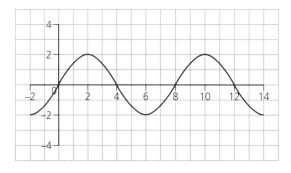

Which line below correctly represents the frequency and amplitude of this wave?

	Frequency/Hz	Amplitude/mm
A	0.125	2
B	8	4
C	125	2
D	12.5	4

[1]

2 An observer detects a sound wave. The observer then notes that the sound wave had its wavelength increased while the velocity of sound remains unchanged. This could be due to:

A the source of the sound moving away from the observer.
B the sound wave moving into air which is less dense.
C the source of the sound moving towards the observer.
D the sound wave moving into air which is more dense.   [1]

3 The refractive index of water is 1.3. Which one of the following diagrams showing the path of a light ray is **not** correct.   [1]

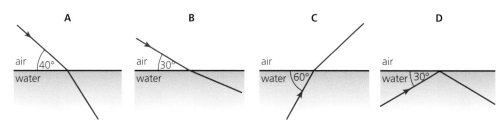

4 Two microwave sources, A and B, are placed 16 cm apart. The sources are in phase with one another, producing microwaves of wavelength 3.2 cm. A detector is placed at point P, which is 272 cm from A. Other distances are given on the diagram.

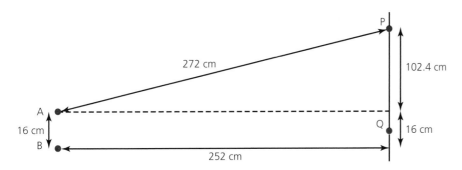

a Calculate
  i the number of wavelengths between A and P
  ii the number of wavelengths between B and P. **[2]**
b The detector is moved along the line PQ. Use your answer to part **a** to describe how the amplitude of the received signal varies as the detector moves from P to Q (the point that is directly opposite to the midpoint of AB). **[4]**

5 A standing wave is set up on a string that is 1.2 m long.

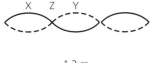

a Explain how a standing wave is formed. **[3]**
b Calculate the wavelength of this standing wave. **[2]**
c Calculate the speed of this wave on the string. The frequency is set to 22.5 Hz. **[2]**
d State whether the point Z is a node or an antinode.
e Discuss how the motion of the points X and Y on this string are
  i similar and
  ii different. **[2]**
f The frequency is increased to 30 Hz. Calculate the new wavelength and redraw the pattern that would now be seen. **[2]**

6 a Explain with the aid of diagrams why transverse waves can be polarised but longitudinal ones cannot be polarised. **[3]**
  b Describe with the aid of a diagram how you could show that light can be polarised. **[3]**

Edexcel sample assessment material 2007

Current and potential difference

Current

An electric current is the rate of flow of charged particles:

$$I = \frac{\Delta Q}{\Delta t}$$

where I is the current and ΔQ is the charge that flows past a point in a time interval Δt.

The SI unit of electric charge is the coulomb, C, and the SI unit of current is the ampere, A:

$$1\,A = 1\,C\,s^{-1}$$

The current direction is conventionally taken to be that of a positive charge, and therefore leaves the positive terminal of a power supply and is directed towards the negative terminal of the power supply. In metals, current is a flow of electrons, which have negative charge. The direction of electron flow is therefore opposite to the direction of the 'conventional current'.

Potential difference

Potential difference, p.d. (or voltage, V), is a measure of energy transfer between two points in an electric circuit. When a charge of 1 C moves through a p.d. of 1 V then the energy transferred is 1 J:

$$V = \frac{W}{Q}$$

where W is the work done (i.e. the energy transferred).

The SI unit of potential difference is the volt, V:

$$1\,V = 1\,J\,C^{-1}$$

Worked Example

In an electric circuit 20 J of heat and light are being transferred by a lamp for every 2 C of charge that passes through it. Calculate the potential difference across the lamp.

Use the above equation to give

$$V = \frac{W}{Q} = \frac{20\,J}{2\,C} = 10\,V$$

Emf

Electromotive force (usually written **emf**) is a measure applied to a source of electrical power such as a battery. It is the energy available per coulomb of charge.

Thinking Task

Explain the difference between potential difference and emf.

Quick Questions

Q1 A current of 2.0 mA flows for 2 minutes. Calculate the charge that has flowed.

Q2 A potential difference of 3 V is measured across a lamp through which there is a current of 1.5 A. Calculate the energy transferred by the lamp in 120 s.

Current-potential difference graphs

The current through a component is measured with an ammeter, connected in series. The potential difference (p.d.) across a component is measured with a voltmeter, connected in parallel. Taking measurements of current and voltage allows a graph of these variables to be plotted.

Measurements of current and p.d. over time may also be taken with current and voltage sensors connected to a computer. This usually allows many more readings to be taken, with greater frequency, and graphs of the readings can be produced by computer software.

The greater the p.d. across a device, the greater the current within it. Some devices, such as a metal at constant temperature, obey **Ohm's law**. This means that the current is directly proportional to the p.d., producing a straight line graph that passes through the origin. If the graph of p.d. against current is not a straight line, then the device is **non-ohmic**, such as a lamp filament.

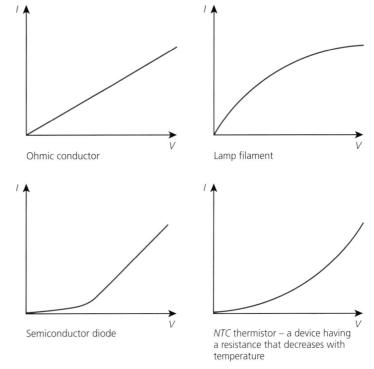

Ohmic conductor

Lamp filament

Semiconductor diode

NTC thermistor – a device having a resistance that decreases with temperature

Ohmic and non-ohmic resistors

Resistance

Resistance R is defined as

$$R = \frac{V}{I}$$

The SI unit of resistance is the ohm, Ω:

$$1\,\Omega = 1\,V\,A^{-1}$$

For a given p.d. across a device, the greater the current, the smaller the resistance of the device. Resistance can be calculated from measurements of current and p.d. The resistance of a device that obeys Ohm's law does not vary with current or p.d.

Worked Example

Find the current through a $20\,\Omega$ resistor when the p.d. across it is $12\,V$.

Rearrange the equation relating resistance, p.d. and current, $R = V/I$, to make I the subject, substitute values and solve:

$$I = \frac{V}{R} = \frac{12\,V}{20\,\Omega} = 0.6\,A$$

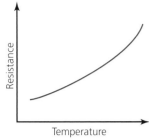

Variation of resistance with temperature in a filament lamp

Resistance and temperature

The resistance of many electrical components changes with temperature. A current flowing through a component can raise its temperature and so change its resistance.

Metals

The resistance of most metals increases with temperature.

- There are free electrons in metals.
- As electrons move through the metal, they collide with vibrations of the lattice and are scattered.
- If the metal gets hotter, the atoms vibrate more vigorously.
- There are then more collisions between the atoms and the electrons.
- The flow of electrons is reduced, i.e. the current decreases.
- If the current decreases, the resistance has increased.

Semiconductors

Semiconductor materials contain far fewer free electrons than metals.

- Semiconductors with a negative temperature coefficient (NTC) have resistance that decreases with temperature. As the material gets hotter, increased vibration of the atoms releases more electrons. This increases the current: resistance has decreased.
- In semiconductors with positive temperature coefficient (PTC), the resistance increases with increasing temperature.

⚙ Thinking Task

Why does a metal lamp filament behave as a non-ohmic conductor?

⁇ Quick Questions

Q1 The current through a lamp is 0.25 A when the p.d. across it is 6.0 V. What is the resistance of the lamp?

Q2 The diagram shows the *V-I* graph for an ohmic conductor. What is the resistance of the conductor?

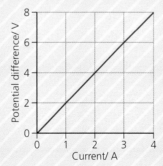

Q3 In 1 minute, 30 C of charge passes through a 12 V lamp. What is the resistance of the lamp?

Power and work in electric circuits

Power

Power is the rate of doing work or the rate of transfer of energy:

$$P = \frac{\Delta W}{\Delta t} = \frac{\Delta E}{\Delta t}$$

The SI unit of power is the watt, W:

$$1\,W = 1\,J\,s^{-1}$$

There are several different equations you can use to work out power in electric circuits.

- From the definition of potential difference:

$$V = \frac{W}{Q} \quad \text{so} \quad W = VQ$$

   giving $$P = \frac{VQ}{t}$$

- From the definition of current:

$$I = \frac{Q}{t} \quad \text{so} \quad Q = It$$

   giving $$P = \frac{VIt}{t} = VI$$

- As $V = IR$ and so $I = V/R$:

$$P = I^2R \quad \text{or} \quad P = \frac{V^2}{R}$$

Work can be calculated from power, or from p.d., current and time:

$$W = Pt \quad \text{so} \quad W = VIt$$

Worked Example

The power dissipated in a lamp is 1.5 W when its resistance is 24 Ω. What is the current in the lamp?

- -

Use the equation relating power, current and resistance:

$$P = I^2R$$

Rearrange to give:

$$I^2 = \frac{P}{R} \quad \text{or} \quad I = \sqrt{\frac{P}{R}}$$

Substitute values and solve:

$$I = \sqrt{\frac{1.5\,W}{24\,\Omega}} = 0.25\,A$$

ResultsPlus
Watch out!

Remember to take the square root.

❓ Quick Questions

Q1 A hairdryer is rated as 230 V, 1.84 kW.
   a What current flows through it?
   b What is its resistance?
Q2 A light bulb transfers 4320 J of energy when left on for 20 minutes. The current in the bulb is 0.3 A. What is the p.d. across it?
Q3 In a physics lab a 6 V electric heater was used to heat a beaker of water. The current in the heater was 2 A and it supplied a total of 3600 J of energy to the water. For how many minutes was the heater switched on?

Resistance and resistivity

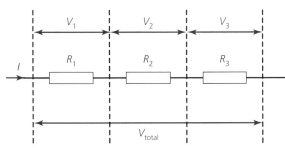

Resistors in series

Resistors in series

At any point in a circuit, the charge flowing in must be the same as the charge flowing out. When resistors are joined in **series**, the current is the same through all of them.

The total energy supplied per coulomb of charge is converted by the resistors. Therefore, the sum of the individual p.d.s across each resistor is equal to the total p.d. applied.

$$V_{total} = V_1 + V_2 + V_3 = IR_1 + IR_2 + IR_3$$

$$R_{total} = \frac{V}{I} = R_1 + R_2 + R_3$$

Resistors in parallel

When resistors are joined in **parallel**, the total current into any junction is equal to the sum of the currents coming out. The p.d. across each resistor is equal to the p.d. applied.

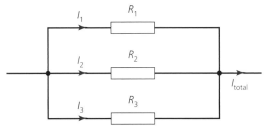

Resistors in parallel

$$I_{total} = I_1 + I_2 + I_3 = \frac{V}{R_1} + \frac{V}{R_2} + \frac{V}{R_3}$$

$$\frac{I_{total}}{V} = \frac{1}{R_{total}} = \frac{1}{R_1} + \frac{1}{R_2} + \frac{1}{R_3}$$

Worked Example

Calculate the resistance marked X. The reading on the ohmmeter is $6\,\Omega$.

Using the series rule, the combined resistance of X and the $12\,\Omega$ resistor is:

$$R = 6\,\Omega - 2\,\Omega = 4\,\Omega$$

Using the parallel rule:

$$\frac{1}{R\Omega} = \frac{1}{4\Omega} = \frac{1}{12\Omega} + \frac{1}{X\Omega}$$

Rearranging and rewriting this gives:

$$\frac{1}{X\Omega} = \frac{1}{4\Omega} - \frac{1}{12\Omega} = \frac{3\Omega - 1\Omega}{12\Omega} = \frac{2\Omega}{12\Omega}$$

so

$$X = \frac{12\Omega}{2\Omega} = 6\,\Omega$$

ResultsPlus
Watch out!

Don't forget to turn the fraction upside down to find X.

Resistivity

Materials have **resistivity** ρ. A specific shape and size of conductor has resistance R:

$$R = \frac{\rho l}{A}$$

where l is the length of the conductor and A is its cross-sectional area.

Worked Example

A 200 cm length of aluminium wire of diameter 1.2 mm has a resistance of 23 Ω. Calculate the resistivity of aluminium.

Calculate the cross-sectional area A:

$$A = \pi r^2 = \pi \times (0.6 \times 10^{-3})^2 = 1.13 \times 10^{-6} \, m^2$$

Calculate the resistivity:

$$\rho = \frac{RA}{l} = \frac{23 \, \Omega \times 1.13 \times 10^{-6} \, m^2}{2.00 \, m} = 13 \times 10^{-6} \, \Omega m$$

ResultsPlus
Watch out!

The units of the length and area must be consistent. These calculations have used metres and metres squared. Don't forgot to divide the diameter by 2 to get the radius.

Number density of conduction electrons

A material with a small value of resistivity will be a good conductor. A good conductor will have a large number of free electrons. The number density n is the number of conduction (free) electrons per unit volume of a material:

$$n = \frac{I}{qvA}$$

where I is the current in a sample, q is the charge on an electron, v is the drift velocity of the free electrons, and A is the cross-sectional area of the sample.

Quick Questions

Q1 Calculate the total resistance and then the current in each resistor for each circuit.

a

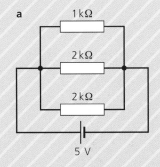

b

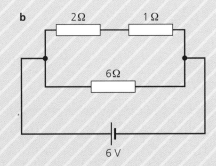

Q2 A copper wire of cross-sectional area $3.2 \times 10^{-6} \, m^2$ carries a current of 0.60 A. Calculate the drift velocity of the conduction electrons. [charge on electron = 1.6×10^{-19} C; number density for copper = $10^{29} \, m^{-3}$]

Q3 The number density of free electrons for copper is $10^{29} \, m^{-3}$ and for silicon is $10^{18} \, m^{-3}$. The resistivity of copper is $10^{-8} \, \Omega m$ and for silicon is $10^3 \, \Omega m$. Comment on the relationship between number density and resistivity.

Thinking Task

A rectangular sheet of conducting paper the same size and thickness as the page you are reading is to be used in an electrical circuit. The resistivity of conducting paper is 35 $\mu\Omega$ m. Take measurements and calculate its resistance if terminals are going to be placed at the top edge and bottom edge (seen as you are reading it) of the sheet.

Potential dividers and internal resistance

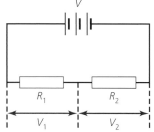

Resistors in series

The potential divider

A potential divider consists of two resistors connected in series with a supply. Because they are in series, the same current flows through both resistors, and the potential difference (p.d.) of the supply is divided between them.

Using Ohm's law:

$$V_1 = IR_1 \quad \text{and} \quad V_2 = IR_2$$

Dividing the first of these equations by the second gives:

$$\frac{V_1}{V_2} = \frac{R_2}{R_2}$$

The circuit can be used as a **potential divider** to provide a p.d. to another device or another circuit that is less than the p.d. of the power supply. If one of the resistors is variable, the p.d. from the potential divider can easily be adjusted.

connections to another device or circuit

A potential divider

Worked Example

What is the p.d. of each resistor in this circuit?

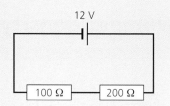

The total p.d. across the two resistors is 12 V. Using the potential divider equation, this is divided up in the ratio of the two resistances:

$$\frac{V_1}{V_2} = \frac{R_1}{R_2} = \frac{100\,\Omega}{200\,\Omega} = \frac{1}{2}$$

$$V_1 = 0.5V_2$$

We also know that the p.d.s add to the p.d. of the supply:

$$V_1 + V_2 = 12\,\text{V}$$

Combine the two equations:

$$0.5V_2 + V_2 = 12\,\text{V}$$

$$1.5V_2 = 12\,\text{V}$$

$$V_2 = 8\,\text{V}$$

Substitute this value into the equation relating V_1 and V_2:

$$V_1 = 0.5V_2 = 4\,\text{V}$$

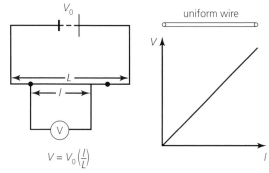

$$V = V_0\left(\frac{l}{L}\right)$$

A potential divider using uniform wire

Potential divider using resistance wire

A potential divider circuit can be made using a single piece of resistance wire in place of the two resistors. This provides a continuously variable p.d. from zero up to the p.d. of the supply.

If a uniform wire is used, the resistance between any two points is proportional to the distance between them. When the wire carries a current, the graph of p.d. between one end and any other point is a straight line.

Internal resistance

The e.m.f. $\mathcal{E}$ of a power supply is the total energy it supplies to each coulomb of charge.

Many power supplies have some internal resistance, so some of the energy supplied to each coulomb is 'lost' due to heating within the power supply.

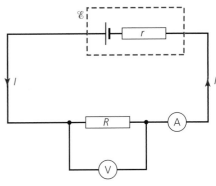

Energy supplied to each coulomb	=	energy transferred by load resistor	+	energy transferred due to internal heating
$\mathcal{E}$	=	IR	+	Ir

How a power supply with internal resistance can be represented in a circuit diagram

so $$\mathcal{E} = V + Ir$$

Terminal potential difference and lost volts

In the above, V is the **terminal potential difference**. Internal resistance is represented by r. The expression Ir is sometimes known as the 'lost volts'.

- An **open circuit** means that the power supply has no connection between its terminals, or is connected to a very high resistance, e.g. a voltmeter. Then $I = 0$ and $V = \mathcal{E}$.
- A **short circuit** means that the power supply terminals are joined by a connection with no resistance. Then $V = 0$ and $I = \mathcal{E}/r$.

Worked Example

A power supply with emf 12.0 V and internal resistance 0.03 Ω operates a motor with a load resistance of 1.5 Ω. What is the current in the circuit? What are the 'lost volts' and what is the terminal p.d.?

Draw a diagram, showing R and r.

Calculate the current:

$$\mathcal{E} = I(R + r)$$

so

$$I = \frac{\mathcal{E}}{(R + r)} = \frac{12\,V}{(1.5+0.03)\,\Omega} = \frac{12\,V}{1.53\,\Omega} = 7.84\,A$$

$R = 1.5\ \Omega$

Calculate the lost volts:

$$\text{lost volts} = Ir = 7.84\,A \times 0.03\,\Omega = 0.24\,\Omega$$

Calculate the terminal p.d.:

$$\text{terminal p.d.} = V = IR = 7.84\,A \times 1.5\,\Omega = 11.76\,V$$

or $\quad V = \mathcal{E} - Ir = 12.0\,V - 0.24\,V = 11.76\,V$

Quick Questions

Q1 A cell of emf 1.5 V and internal resistance 0.30 Ω is connected across a 2.7 Ω load resistor.
  a What is the current in the circuit?
  b What are the 'lost volts'?
  c What is the terminal p.d.?

Q2 A battery has an emf of 3.0 V and an internal resistance of 2.0 Ω. Two resistors, of resistance 4.0 Ω and 6.0 Ω, are connected in series to the battery.
  a What is the current in the circuit?
  b What is the potential difference across the 4.0 Ω resistor?
  c What is the potential difference across the 6.0 Ω resistor?
  d What are the 'lost volts'?

Thinking Task

Why do the headlights on a car dim when the starter motor is operated? Illustrate your answer using the following data: battery emf = 12 V, battery internal resistance = 0.040 Ω, starter motor current = 100 A.

DC electricity checklist

By the end of this section you should be able to:

Revision spread	Checkpoints	Spec. point	Revised	Practice exam questions
Current and potential difference	Describe electric current as the rate of flow of charged particles and use the expression $I = \Delta Q/\Delta t$	50	☐	☐
	Define and use the concept of emf	59	☐	☐
	Use the expression $V = W/Q$	51	☐	☐
Current-potential difference graphs	Interpret current/potential difference graphs, including those for non-ohmic materials	56	☐	☐
	Understand how ICT may be used to obtain current/potential difference graphs and compare this with traditional techniques	55	☐	☐
	Use the fact that resistance is defined by $R = V/I$ and that Ohm's law is a special case when $I \propto V$	54	☐	☐
	Investigate and recall that the resistance of metallic conductors increases with increasing temperature and that the resistance of negative temperature coefficient thermistors decreases with increasing temperature	60	☐	☐
	Explain how changes of resistance with temperature may be modelled in terms of lattice vibrations and number of conduction electrons	62	☐	☐
Power and work in electric circuits	Investigate and use the expressions $P = VI$ and $W = VIt$	53	☐	☐
	Recognise and use related expressions, e.g. $P = I^2R$ and $P = V^2/R$	53	☐	☐
Resistance and resistivity	Recognise and use the relationships between current, voltage and resistance, for series and parallel circuits	52	☐	☐
	Know that these relationships are a consequence of the conservation of charge and energy	52	☐	☐
	Investigate and use the relationship $R = \rho l/A$	57	☐	☐
	Use $I = nqvA$ to explain the large range of resistivities of different materials	61	☐	☐
Potential dividers and internal resistance	Investigate and explain how the potential along a uniform current-carrying wire varies with the distance along it and how this variation can be made use of in a potential divider	58	☐	☐
	Define and use the concepts of emf and internal resistance and distinguish between emf and terminal potential difference	59	☐	☐

ResultsPlus
Build Better Answers

A 3 V battery, two identical lamps marked 1.5 V and a variable resistor are connected in this circuit.

a The variable resistor initially has a resistance much larger than either lamp. Explain why each lamp will be equally bright. **[3]**

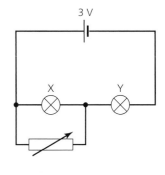

Student answer	Examiner comments
■ There is very little current in the variable resistor, so each lamp has roughly the same current.	■ This **basic answer** shows some appreciation of why the current in each lamp is the same.
▲ There is very little current in the variable resistor, so each lamp has roughly the same current. Each lamp will divide the potential difference equally (1.5 V each) so the power (= IV) for each lamp is the same.	▲ This **excellent answer** has considered V and I, both of which determine whether lamps are bright/dim. It would gain all 3 marks.

b The resistance of the variable resistor is reduced until it is much less than the resistance of either lamp. Explain what happens to the brightness of each lamp. **[4]**

Student answer	Examiner comments
■ The current in lamp Y will mainly split into the variable resistor, with very little current in X. X will go off and Y remains bright.	■ This **basic answer** appreciates what happens to current, but does not go far enough in the explanation.
▲ The current in lamp Y will mainly split into the variable resistor, with very little current in X. The potential difference (= IR) will be very small across lamp X. The potential difference across Y will be close to 3 V. Lamp X goes off while lamp Y becomes very bright.	▲ This **excellent answer** demonstrates an understanding of current and an ability to apply a relevant equation ($V = IR$) to the situation. The student also appreciates that both I and V need to be considered to comment on brightness. It would gain all 4 marks.

Practice exam questions

1 Two pieces of the same semiconducting material are joined in series to a battery. One piece X has twice the cross-sectional area of the other piece Y. The ratio

$$\frac{\text{speed of conducting charges in X}}{\text{speed of conducting charges in Y}}$$

is equal to:
A 1
B 2
C $\frac{1}{2}$
D 4

2 A cube of metal has sides of length x. The electrical resistance between opposite faces of the cube is:
A directly proportional to x.
B proportional to x^2.
C inversely proportional to x.
D independent of x.

3 A student investigates the use of rechargeable cells in a torch. The diagram shows the circuit used to obtain readings of voltage and current for the torch bulb. The results obtained are voltage = 2.68 V, current = 0.31 A.

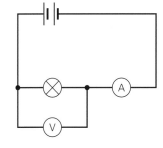

a i Calculate the resistance of the torch bulb. **[2]**
 ii The cells form a battery with an emf of 2.80 V. Show that the internal resistance of the battery is about 0.4 Ω. **[2]**
b Each cell is labelled: 'capacity 2 A h' (A h is amp hours).
 i Show that this corresponds to a total stored charge of about 14 000 C for the battery of two cells. **[2]**
 ii Calculate the time for which the battery could maintain a current of 0.31 A in the torch bulb. **[2]**
c The internal resistance of the cells increases as they are used. Explain what effect this will have on the efficiency of the system. **[2]**

Edexcel June 2007 PSA1

Photoelectric effect

Evidence of the particle nature of light comes from the photoelectric effect. A photon is a package (quantum) of electromagnetic energy. When a single photon is absorbed by a metal surface, its energy is transferred to a single electron, which may then be released from the metal. This process is called **photoelectric emission** and the released electron is known as a **photoelectron**.

The work function

The minimum energy required to release the photoelectron from the metal surface is called the **work function**, ϕ, of the metal. For photoelectric emission to occur, the energy of the photon must be equal to or greater than the work function.

If the photon's energy, E, is *just* enough to release a photoelectron, then its frequency is called the **threshold frequency**, f_0:

$$E = hf_0 = \phi$$

where h is the Planck constant (6.63×10^{-34} J s).

If the energy of the photon is greater than the work function, then the photoelectron can acquire some kinetic energy. By *energy conservation*:

$$\text{photon energy} = \begin{array}{c}\text{work done in releasing}\\ \text{the electron}\end{array} + \begin{array}{c}\text{kinetic energy}\\ \text{of electron}\end{array}$$

This can be written as

$$hf = \phi + E_{k\,max}$$

or

$$hf = hf_0 + E_{k\,max}$$

or

$$hf = \phi + \tfrac{1}{2}mv^2_{max}$$

where $E_{k\,max}$ ($= \frac{1}{2}mv^2_{max}$) is the maximum kinetic energy of the photoelectron.

Photoelectrons may have less than the maximum kinetic energy if they transfer energy to the metal on their way to the surface.

If the photon energy is *less* than the work function, the energy absorbed by the metal just causes a slight amount of heating.

Results
Watch

Convert nm to m using
$1\,\text{nm} = 1 \times 10^{-9}\,\text{m}$.

> ### Worked Example
>
> Antimony–caesium has a threshold wavelength of 700 nm. What is its work function in joules?
>
> --
>
> Using the equations $\phi = hf_0$ and $c = f_0\lambda_0$ and substituting in the resulting equation for ϕ gives:
>
> $$\phi = \frac{hc}{\lambda_0} = \frac{6.63 \times 10^{-34}\,\text{J s} \times 3.00 \times 10^8\,\text{m s}^{-1}}{700 \times 10^{-9}\,\text{m}} = 2.84 \times 10^{-19}\,\text{J}$$

The wave model and the photon model

Under certain conditions, light diffracts. Diffraction can only be explained using a wave model.

In the photoelectric effect, a wave model of light predicts that weak radiation would eventually release large numbers of electrons with low energy, but this is not what happens.

The photoelectric effect can only be explained using a photon model. Weak radiation causes instantaneous release of some electrons if the frequency is greater than the threshold frequency. The kinetic energy of individual photoelectrons depends only on the frequency of the radiation, not its intensity.

Stopping voltage

If the metal surface is connected to a positive potential, the photoelectron is attracted back to it. To escape from the surface, the kinetic energy of the photoelectron is used to do work against the electrostatic force. If the potential is increased, eventually even the most energetic electrons fail to escape, and the potential is called the **stopping voltage**, V_s. The charge on each electron is e. So

$$\tfrac{1}{2}mv^2_{max} = eV_s$$

$$hf = \phi + eV_s$$

Hence

$$hf = hf_0 + eV_s$$

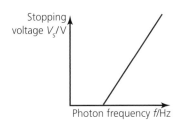

This can be rearranged in the form of the general equation of a straight line, $y = mx + c$:

$$V_s = \left(\frac{h}{e}\right)f - \left(\frac{hf_0}{e}\right)$$

So a graph of V_s (y-axis) against f (x-axis) will have gradient h/e and intercept $-hf_0/e$.

The electronvolt

When dealing with small energies, such as in the photoelectric effect, it is convenient to use a smaller unit than the joule. We use the electronvolt (eV), which is the energy gained when an electron moves through a potential difference of 1 V. The charge on an electron is $e = 1.6 \times 10^{-19}$ C. So

$$1\,\text{eV} = 1.6 \times 10^{-19}\,\text{J}$$

(?) Quick Questions

Q1 The work function of sodium is 3.78×10^{-19} J. What is its threshold frequency?

Q2 Describe and explain what will happen if light with a frequency less than the threshold frequency shines on sodium.

Q3 Blue light of frequency 7.06×10^{14} Hz shines on sodium. Calculate the maximum energy of the photoelectrons released.

Thinking Task

How would the graph of stopping voltage against frequency be different for a metal with a smaller work function?

Photons, spectra and energy levels

Photon energy

The behaviour of light can be modelled as a wave or as a photon (particle). The two models are linked by the relationship:

$$E = hf$$

where E is the energy of the photon (J), h is the Planck constant (6.63×10^{-34} J s) and f is the wave frequency (Hz).

ResultsPlus
Watch out!

Convert nm to m using
$1\,\text{nm} = 1 \times 10^{-9}\,\text{m}$.

Worked Example

Light of a certain orange colour has a wavelength of 589 nm. What is the energy of one photon of this light? Speed of light $c = 3.00 \times 10^8\,\text{m s}^{-1}$.

Combine the equations $E = hf$ and $f = c/\lambda$ relating energy, frequency, speed and wavelength, and then substitute values in and solve the resulting equation:

$$E = \frac{hc}{\lambda} = \frac{6.63 \times 10^{-34}\,\text{J s} \times 3.00 \times 10^8\,\text{m s}^{-1}}{589 \times 10^{-9}\,\text{m}} = 3.38 \times 10^{-19}\,\text{J}$$

Spectra and energy levels

When light is spread out into its separate colours, this is called a spectrum. The light from a tungsten filament lamp produces a continuous spectrum. If the light is from a gas discharge tube, we observe a line spectrum when the light is passed through a slit, i.e. only certain lines of distinct colour are present, separated by gaps. These spectra are due to the emission of energy, and so are called **emission spectra**. If light from a tungsten filament lamp is passed through sodium vapour, the spectrum produced has black lines in it. This is called an **absorption spectrum**.

Line spectra can be explained using the photon model. Each coloured line in an emission spectrum arises when an electron within an atom drops from one **energy level** to another and a photon is emitted. The dark lines in an absorption spectrum arise because an electron absorbs energy and moves from a lower to a higher energy level.

The energies of the photons emitted or absorbed correspond to the differences between energy levels in the atom. The electrons of an atom are only allowed to have certain values of energy. Each atom has its own unique set of discrete (separate) electronic energy levels and hence its own unique line spectrum.

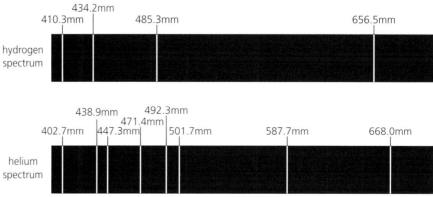

Emission spectra for hydrogen and helium

The photon energies can be calculated from measurements of the wavelengths emitted or absorbed. An energy level diagram is like a graph with energy increasing upwards. A downward-pointing arrow represents an electron losing energy and giving out a photon. The figure shows some of the transitions for a hydrogen atom.

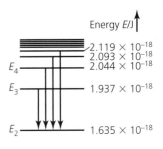

Worked Example

Calculate the frequency of light emitted by a transition from level E_3 to level E_2 in the energy level diagram on the right.

Find the difference in energy between levels:

$$E_3 = 1.937 \times 10^{-18}\,\text{J}, \quad E_2 = 1.635 \times 10^{-18}\,\text{J}$$

The energy lost by the electron is $E_3 - E_2$, so photon energy is:

$$E = E_3 - E_2 = (1.937 - 1.635) \times 10^{-18}\,\text{J} = 3.02 \times 10^{-19}\,\text{J}$$

Calculate the frequency from $f = E/h$:

$$f = \frac{E}{h} = \frac{3.02 \times 10^{-19}\,\text{J}}{6.63 \times 10^{-34}\,\text{J s}} = 4.56 \times 10^{14}\,\text{Hz}$$

Some of the transitions for a hydrogen atom

ResultsPlus
Watch out!

Be careful with the powers of 10.

Maximum energy changes

Because $E = hf$, the highest frequency of radiation that could be emitted corresponds to the most energetic photon that could be released. This occurs when an electron moves from a state where the atom has the most energy to one where it has the least, known as the **ground state**.

The same amount of energy being absorbed corresponds to an atom in the ground state changing to a state where it has the most energy. This occurs when an electron is given sufficient energy to escape completely from the atom, resulting in ionisation of the atom.

(?) Quick Questions

Q1 Calculate the wavelength of light emitted by a transition from level E_4 to level E_2 in the energy level diagram above.
Q2 Explain how emission and absorption spectra are produced.
Q3 Explain why the line spectrum from a chemical element can be used to identify that element.

Thinking Task

Photon energy increases as frequency increases. Explain how photon energy varies with wavelength.

Radiation flux and solar cells

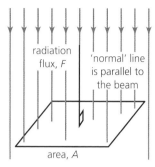

radiation flux, *F*

'normal' line is parallel to the beam

area, *A*

Radiation flux

Radiation flux and solar cells

Solar cells transform light energy into electrical energy. The rate at which a beam of light supplies energy to a particular area is called the **radiation flux** (or intensity) of the beam.

Radiation flux

Radiation flux is given by the relationship

$$F = \frac{P}{A}$$

where F is the radiation flux (W m^{-2}), P is the power (W) and A is the area perpendicular to the beam (m^2).

Power is the rate of supply of energy, so

$$P = \frac{E}{t}$$

where E is the energy (J) and t is the time (s).

Worked Example

A light beam of solar flux 10 W m^{-2} is incident on a solar cell of area 4 cm^2. Calculate the maximum energy transformed by the cell in 1 minute.

Use the equations relating radiation flux, power, area, energy and time:

$$F = \frac{P}{A} \quad \text{and} \quad P = \frac{E}{t} \quad \text{to give} \quad F = \frac{E}{At}$$

Rearrange this to make E the subject, substitute values and solve:

$$E = FAt = 10\,\text{W m}^{-2} \times 4 \times 10^{-4}\,\text{m}^2 \times 60\,\text{s}$$
$$E = 0.24\,\text{J}$$

Efficiency

A solar cell will not transform all the light energy to electrical energy because it is not perfectly efficient.

For any process or device:

$$\text{efficiency} = \frac{\text{useful energy output}}{\text{total energy input}}$$

or

$$\text{efficiency} = \frac{\text{useful power output}}{\text{total power input}}$$

Efficiency is often expressed as a percentage, worked out by multiplying the efficiency calculated above by 100.

Worked Example

A solar-powered vehicle has a total area of $8\,m^2$ covered with solar cells that are 25% efficient. If the solar flux is $600\,W\,m^{-2}$, what is the useful power output from the solar cells?

Calculate power input:

$$\text{power input} = 600\,W\,m^{-2} \times 8\,m^2 = 4800\,W$$

Use the efficiency to calculate power output. The quoted efficiency = 25% = 0.25, so

$$\text{useful power output} = 0.25 \times 4800\,W = 1200\,W$$

Solar cells

Governments across the world are looking for viable ways of providing energy, without the use of fossil fuels. Solar cells provide electricity from a free, renewable source. Scientific knowledge may be used by society to make decisions about the viability of solar cells as a replacement for other energy sources.

Considerations include:

- efficiency of the cells
- cost
- availability of materials
- global energy prices
- impact of continuing use of fossil fuels.

At present, solar cells are of limited use as an energy source. They are often used for small, portable devices such as watches and calculators. Solar cells may be useful in countries that have guaranteed sunshine for most of the year. But until more efficient cells are developed, they are not suitable for places with little sunshine in the winter. The development of more efficient cells is expensive, and industry is not likely to make this investment unless governments also invest in the technology.

? Quick Questions

Assume that the radiation flux from the Sun is $1000\,W\,m^{-2}$ at the Earth's surface.

Q1 When sunlight shines on a solar cell of area $0.25\,m^2$, the cell produces an output power of $40\,W$. What is the percentage efficiency of the cell?

Q2 A solar cell array consists of 40 photovoltaic cells. Each cell has an area of $10\,cm^2$ and has an efficiency of 8%. Calculate the power output of the array.

Q3 A beam of sunlight strikes the surface of a photovoltaic cell. The surface area of the cell is $0.020\,m^2$. Each photon has an energy of $4.0 \times 10^{-19}\,J$.
  a How much energy is delivered to the cell each second?
  b How many photons arrive at the cell each second?

⚙ Thinking Task

A satellite in orbit needs approximately $4\,kW$ of power.
a What area of solar cells does it need?
b State any assumptions you made when working out your answer, and why you think these are reasonable assumptions.

Nature of light checklist

By the end of this section you should be able to:

Revision spread	Checkpoints	Spec. point	Revised	Practice exam questions
Photoelectric effect	Explain how the behaviour of light can be described in terms of waves and photons	63		
	Recall that the absorption of a photon can result in the emission of a photoelectron	64		
	Understand and use the terms threshold frequency and work function	65		
	Recognise and use the expression $hf = \phi + \frac{1}{2}mv^2_{max}$	65		
	Use the non-SI unit, the electronvolt (eV) to express small energies	66		
	Explain how wave and photon models have contributed to the understanding of the nature of light	71		
Photons, spectra and energy levels	Recognise and use the expression $E = hf$ to calculate the highest frequency of radiation that could be emitted in a transition across a known energy band gap or between known energy levels	67		
	Explain atomic line spectra in terms of transitions between discrete energy levels	68		
Radiation flux and solar cells	Define and use radiation flux as power per unit area	69		
	Recognise and use the expression efficiency = $\dfrac{\text{useful energy (or power) output}}{\text{total energy (or power) input}}$	70		
	Explore how science is used by society to make decisions, for example about the viability of solar cells as a replacement for other energy sources	72		

ResultsPlus
Build Better Answers

The diagram shows monochromatic light falling on a photocell.
a Explain the following observation: Initially there is a current which is measured by the microammeter. As the potential difference is increased the current reading on the microammeter decreases.

[4]

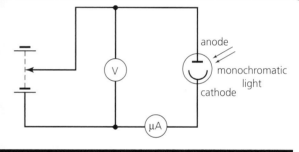

Student answer	Examiner comments
■ Photoelectrons go from the cathode to the anode. As the potential difference is increased it stops some of these electrons from reaching the anode as the anode is connected to the negative terminal of the battery.	■ This **basic answer** recognises some key facts. It recognises the direction of photoelectrons but has neglected to explain why they are there. It also recognises that there is a reverse potential difference preventing some of the electrons reaching the anode.
▲ The photon energy of the light is greater than the work function of the material in the cathode, so it releases electrons from the cathode. As the potential difference is increased, it stops more of these electrons from reaching the anode, as the anode is connected to the negative terminal of the battery. The electrons have a range of energies, so some will still get to the anode, until the potential difference is increased sufficiently to stop all of them.	▲ This **excellent answer** adds the missing details, and would gain full marks.

b When this current is zero the value of potential difference is known as the stopping voltage. Explain whether this stopping voltage will change if the wavelength of the monochromatic light is decreased without changing its intensity. **[3]**

Edexcel January 2007 Unit Test 4

Student answer	Examiner comments
■ This voltage will need to increase because the electrons will have more energy.	■ In this **basic answer**, the student has recognised the relationship between the wavelength of light and delivering more energy, but hasn't explicitly stated this fact.
▲ The light will consist of photons of larger energy. This will release photoelectrons, which will have a larger maximum kinetic energy. These will require a larger stopping voltage to remove their kinetic energy and reduce the ammeter reading to zero.	▲ This **excellent answer** explains the relationship between photon size, the kinetic energy of the photoelectrons and the required stopping voltage. It would be awarded the 3 marks.

Practice exam questions

1 The units of radiation flux could be

    A W
    B Wm^{-2}
    C J
    D Jm^{-2}

2 A fluorescent lamp consists of a glass tube containing a small amount of mercury vapour. When the lamp is switched on, the mercury atoms emit photons of ultraviolet (UV) radiation. A phosphor coating inside the tube converts this radiation into visible light.

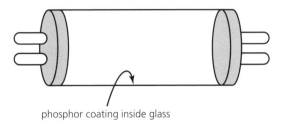

phosphor coating inside glass

When a UV photon hits the coating it excites an electron into a higher energy level. As the electron falls back down, it emits a photon of visible light.

    a Copy the following electron energy level diagram for an atom of the phosphor coating. Add labelled arrows to your diagram to illustrate this process. Start with the absorption of a UV photon and end with the emission of a photon of visible light. **[2]**

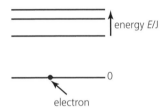

    b UV radiation with wavelengths in the range 320 nm–400 nm is emitted by the mercury atoms. Show that the emitted photons have a minimum energy of about 3 eV. [1 nm = 1×10^{-9} m] **[4]**

Edexcel January 2007 PSA2

3 a Use a diagram to show what is meant by diffraction. **[2]**
    b X-ray diffraction images such as the one shown here led scientists to the first understanding of the structure of DNA.
       The image shown is a negative on photographic film. The dark bands correspond to maximum X-ray detection. Use diagrams to explain how two X-ray waves overlap to produce maximum intensity. **[2]**
    c State two pieces of information about the structure of DNA that can be deduced from this diffraction image. **[2]**
    d Electrons can also be used to produce diffraction patterns and hence to study materials in this way. What does this tell you about the behaviour of electrons when passing through such materials? **[2]**

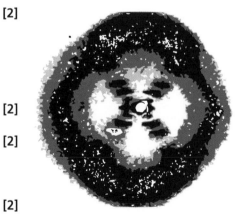

Edexcel January 2008 PSA2

Unit 2: Practice unit test

Section A

1 The flashgun of a camera operates for 1 ms. During this time, 0.05 C of charge flows. The current is

    A 5×10^{-8} A
    B 5×10^{-5} A
    C 0.02 A
    D 50 A   **[1]**

2 A current of 0.2 A flows in a copper wire of cross-sectional area 2.0×10^{-7} m^2. The number density of free electrons for copper is 1.0×10^{29} m^{-3}. What is the drift velocity, in m s^{-1}, of the electrons?

    A 6×10^{-5}
    B 2×10^{-42}
    C 2×10^{4}
    D 600   **[1]**

3 A valid set of units for radiation flux is

    A W
    B J m^{-2}
    C W m^{-2}
    D J   **[1]**

4 The smallest distance between two points on a progressive wave that have a phase difference of 60° is 0.05 m. If the frequency of the wave is 500 Hz, the speed of the wave, in m s^{-1}, is

    A 25
    B 75
    C 150
    D 1666   **[1]**

5 A light beam directed onto a thin slit can produce a fuzzy outline of the slit on a screen. This is due to

    A refraction
    B diffraction
    C polarisation
    D reflection   **[1]**

6 The effective resistance between P and Q in the diagram is 12 Ω. What is the value of X, in ohms?

    A 9
    B 12
    C 15
    D 36   **[1]**

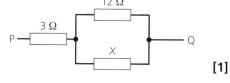

7 Two resistors are connected in parallel to a battery. The power dissipated by the 100 Ω resistor is X watts.

The power, in watts, dissipated by the 200 Ω resistor is

    A $X/4$
    B $X/2$
    C X
    D $2X$ **[1]**

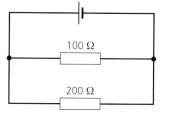

[7 marks]

Section B

8 a The refractive index for red light is 1.3. Calculate the angle of refraction when a ray of red light enters the raindrop at an angle of incidence of 27°. **[2]**

b Calculate the critical angle for red light in a raindrop. **[2]**

c A ray of red light hits the back of a raindrop. Take a suitable measurement and then copy and complete the diagram to show what will happen to this ray of light. **[4]**

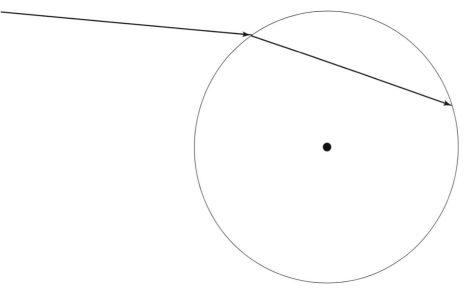

Edexcel June 2008 PSA2

9 a Explain what is meant by plane polarised light. **[1]**

b Explain how you would measure the rotation of the plane of polarisation of light having passed through a solution of water and sugar. **[3]**

10 The image on the right is produced when electrons are targeted on a crystal.

a What conclusions can you draw about the atoms within the crystal?

b What conclusion can you draw about the behaviour of the electron beam? **[1]**

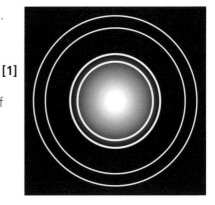

11 A group of students is discussing why the resistance of the metal filament of a lamp and the resistance of an NTC thermistor respond differently to changes in termperature.

One student says that the increased vibrations of the atoms affect the conduction process.

Another student says that as the temperature increases more electrons can break free of the atoms and take part in conduction.

Both students are correct. Explain how these **two** effects apply to the lamp and the thermistor. **[5]**

Edexcel sample assessment material 2008

12 A car is parked with its headlights on. The owner gets in and presses the ignition switch. The headlights dim while the starter motor is trying to turn the engine over. Once the engine is running, the headlights return to their normal brightness.

Explain why the headlights dim in this situation, using the fact that all batteries have some internal resistance. **[4]**

13 Three identical 1.5 V lamps are connected to a 3 V battery, as shown in the diagram. Explain what will happen to the brightness of each bulb when the switch is closed. **[5]**

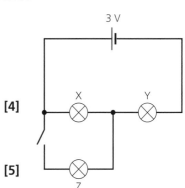

14 Sodium street lamps are usually red when first switched on in the evening. After several minutes they take on their normal yellow colour.

In these lamps, light is emitted as a current passes through the sodium vapour. However, when the lamp is first switched on, the sodium is solid, so little vapour is present until it warms up. The red colour in the first few minutes is due to neon gas, which the lamp also contains.

a Explain how atoms in a vapour emit light. **[2]**
b The light appears yellow because the spectrum of sodium is dominated by two lines with wavelengths of 589.0×10^{-9} m and 589.6×10^{-9} m, respectively.
   i Explain what is meant by a spectral line. **[1]**
   ii Calculate the frequency of the light with a wavelength of 589.0×10^{-9} m. **[2]**
c The light emitted by neon vapour appears red. Explain why atoms of different elements produce light of different colours. **[3]**
d Light is a transverse wave. Explain the meaning of transverse. **[1]**

Edexcel January 2007 PSA1

15 According to a study carried out by Bell Labs in New Jersey, USA, the signal from a mobile phone could be used to detect your pulse and breathing rate. The phone does not need to be answered and so could be used to locate unconscious survivors of earthquakes.

When the phone rings, microwaves are emitted, which then reflect back to the phone from different parts of the owner's body, such as the chest, heart and lungs.

The diagram opposite shows some of these reflected waves being displayed on the screen of an oscilloscope.

Time/μs⟶

a Explain why the microwaves are reflected off different parts of the body. **[1]**
b Give two reasons why the amplitudes of the peaks vary. **[2]**
c Give one reason why the time between the peaks varies. **[1]**
d If the microwaves are reflected from a moving object, such as a beating heart, a Doppler shift is observed.
   i Explain what is meant by the term Doppler shift and how it occurs.

[2]

   ii Describe and explain the observed changes in the wavelength and the frequency of the detected microwave signal when the heart is contracting (moving away from the microwave source). **[3]**

16 Babies' food sometime carries the following warning:

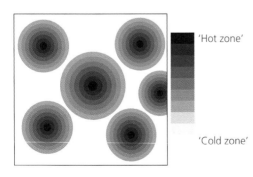

'Do not warm feeds in a microwave oven as this may cause uneven heating and could scald your baby's mouth.'

An Internet site gives the following explanation:

Coherent microwaves are emitted in all directions from a source within the oven. The waves reflect off the metal walls so that the microwave radiation reaching any particular point arrives from several different directions. The waves interfere and set up standing waves. This produces the pattern of hot and cold zones observed in food heated in a microwave oven.

a Explain the meanings of the following words from the passage:
   i coherent **[1]**
   ii standing wave **[2]**
b Copy the diagram above, and on your copy mark a possible position of one antinode, and label it A. **[1]**

c   The frequency of the radiation used in a microwave oven is 2.45×10^9 Hz.
    Show that the wavelength of the microwave radiation is about 12 cm.   **[1]**
d   The diagram below shows two different paths by which microwaves can
    reach the point X. Find the path difference for waves reaching point X by
    the paths shown.   **[1]**

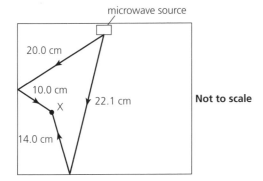

microwave source

20.0 cm

10.0 cm

X   22.1 cm   **Not to scale**

14.0 cm

e   Assuming waves do not reach point X along any other path, explain
    whether you would expect this point to be a microwave node or antinode.   **[3]**
f   Some microwave ovens use two separate microwave frequencies to
    overcome the problem of uneven heating. Explain how this helps.   **[2]**

17 To preserve food effectively, packaging film must isolate food from bacteria in
    the environment. The film is thoroughly inspected for holes, but many inspection
    techniques can damage it.

    A new technique has been developed to check the film. The film is passed between
    a roller electrode and a conducting liquid, in which a second electrode is placed,
    as shown below. The resistance between the two electrodes is measured; even a
    pinhole in the film can greatly reduce this resistance by filling with the liquid and
    creating a conducting path.

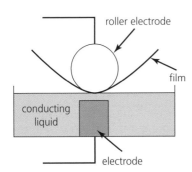

roller electrode

film

conducting
liquid

electrode

a   Copy and complete the diagram above showing a circuit you could
    use to measure this resistance.   **[2]**
b   For undamaged film the resistance measured between the electrodes is very
    high. Show that when a circular hole of diameter 1.0×10^{-4} m is present and
    fills with the conducting liquid the resistance measured is about 170 Ω. Assume
    that the resistance through the liquid between the lower electrode and the film
    is so small it can be ignored.   **[3]**
    [resistivity of conducting liquid = 2.7×10^{-3} Ω m, film thickness = 5.0×10^{-4} m]

[63 marks]
[Total 70 marks]

Exploring physics

Introduction

All AS students are required to carry out one piece of assessed practical work. This work should be based on either a case study or a visit that involves an application of physics. It will help if you have a copy of the marking grids with you as you read this advice.

This section explains how the visit/case study and the practical work are organised, but it is really just a development of the 'How Science Works' themes that you covered at GCSE. However, unlike GCSE, your teachers are not allowed to mark drafts of your work to help you to improve it.

You will be assessed on the following:

- a summary of your visit *or* your case study
- the plan for your practical work
- how you carry out your plan and measure your variables
- the analysis of your practical work and the conclusions you draw.

The visit

Your teacher will have organised the trip for you (or for the whole class). You may have received some briefing material prepared by your teacher, or by people at the place you are going to visit. You will get the most out of your trip if you read this before you go.

You are going to look at some physics in action and write a report on what you see, so a notebook might be a good idea. You should have a copy of the marking grids too.

Just by going you get mark S1. Start your report by stating where you went (mark S2) and briefly describing what you saw (mark S3). Remember, it is the *physics* you should be describing, so if you are visiting a manufacturer of composite materials, for mark S4 you should describe how the materials will stand up to the design loads. You should explain the physics you are seeing using the correct terminology (in this case, explain the Young modulus for marks S4 and S5), and then for mark S7 describe how knowledge of this physics helps the engineer design the parts.

In your report, you should use some of the specialist terms that you heard or read about – make sure you ask your guide about these. While there, find out one piece of technical data that you can use in your report to help bring it to life and to get mark S6 – a diagram helps.

You should describe briefly how the physics helps improve the design or makes the task easier or safer (mark S8).

The case study

You get mark S1 just by going to the library and using three different sources of information. This is a research project, so you must provide the full details of the sources (mark S2). The Internet is only one type, no matter how many different websites you quote. A library might help, but so will people – try asking. You will have been given a briefing paper, so you will have a good idea what area you are looking at. As with the visit, you are reporting on how physics helps people to improve things – for example, you might consider how knowledge of materials improves either fishing rods or reading glasses.

The case study marks for S3 to S8 can be got in the same way as for the visit (see above).

The summary

Your report (of your visit *or* your case study) must be structured with sub-headings (mark R2), so it is a good idea to start with these. You can produce your report using

Thinking Task

Write down a list of sub-headings you could use for a report on a visit. The assessment criteria should help.

a computer, so why not get something you can be proud of? Certainly, grammar and spelling should be checked (mark R1), and you could also use images from a website. Finish with a link to the practical work. Dont forget to credit all your sources.

Check the marking grids to make sure you have included everything you can.

ResultsPlus
Examiner tip

Stick to 600 words maximum.

Planning the practical work

From now on, all the work you do is done in the lab, and you may not take anything outside. But you will be given a list of the marking grids, so you will have some guidance, and your teacher can answer questions that don't actually give you marks.

You will be doing practical work that is on familiar topics, so thinking about the practical work you have already done will be helpful. Before you start, get clear in your mind what you are trying to measure. Keep thinking of this as you go through the work, and check the marking grids to make sure you have included everything you need to in your plan, which should be well organised and methodical (mark P14).

Headings
You will need to structure your report, so start with your headings. You might use these:

- Aim
- Apparatus
- Method
- Measurements
- Analysis
- Evaluation
- Conclusion

ResultsPlus
Examiner tip

The better your plan, the better your experiment.

You will want to list all your materials first (mark P1), but why not put the diagram (mark P13) here too? It will help you to think clearly. You could do a quick sketch on a spare piece of paper first, but you should spend a little time and care on your final diagram. Use a ruler and sharp pencil, and label dimensions accurately.

Variables
What are your two variables? You need to say which is the independent variable and which is the dependent one (mark P7). You need to find out how much they will vary and decide how many readings you will take. Remember, you will be plotting a graph, so you will need at least six data points – will you have time for more? To do this, you must think how you will be efficient in your method – you do not have unlimited time.

You must also be clear what you will do with your readings (mark P11). For instance, if you are finding the resistivity of a wire, you will probably plot resistance against length, but will you need to measure the diameter of the wire too. If you are going to bend a composite material, then you must decide what you will need to do to find the Young modulus. So you will have to take readings of quantities other than your two variables. You should also state what graph you will plot – see the graph section on pages 84–86.

ResultsPlus
Watch out!

You can use a multimeter to measure the resistance of a piece of wire, but not that of a light bulb when it is on. Can you explain why?

You need to control all the other variables, but don't forget to say what they are and how you will do it (mark P8).

Quick Questions

Q1 In an experiment, what is:
  a the independent variable?
  b the dependent variable?
Q2 Think of a resistivity experiment, and name a control variable.

Instruments

Notice that you must choose your measuring instruments for two quantities and justify your choices – this is for *4 marks* (marks P2–P5), so you must get this right. These quantities do not both have to be your variables, but one probably will be. You need to think about precision, accuracy and perhaps sensitivity at this stage. You must say something about how you will use your instruments (mark P6). A wire's diameter must be measured using a right-angled pair of readings at different distances down the wire. Clearly, this is repeating your readings, and (for mark P9) you must say why you will – or will not – repeat your readings. If you are heating a liquid, you cannot repeat readings because the temperature has changed. You can repeat the experiment but not the readings, so how do you plan for accuracy here?

Range

Plan to spread your readings over as wide a range as seems sensible (mark M4), but keep the physics in mind. For example, if you are finding '*g* by free fall', you might think about dropping the object on to the floor, but if you go much further you will have air resistance to consider.

Safety

ResultsPlus ✓
Examiner tip

Don't try to make up silly safety precautions you won't use anyway, 12 V is low voltage, which is safe unless you short out the supply and produce a very large current.

Your experiment is probably pretty safe already, so you should say *why it is safe* (mark P10). It is good practice to do a 'risk assessment' for your experiment – you should be familiar with these from previous practical work.

Uncertainties

You need to identify the main sources of uncertainty in your results (mark P12). *Random errors* occur because what you are measuring is not uniform, or because your repeat reading isn't exactly the same. *Systematic errors* are caused by things such as faulty instruments or inadequate apparatus. You can often do something about these, such as checking your instruments for zero errors before you start.

Taking the measurements

ResultsPlus ✓
Examiner tip

Make sure you use SI units in your calculations – so when using a micrometer, you must convert the reading to metres.

Bigger values for variables are easier to read and more likely to be accurate.

Draw a table ready for your results before you start. As you start, think again about how many readings you want to take to give you a reliable conclusion (mark M3). Remember, you will want at least six points on your graph. If you are not taking repeat readings, you should take enough of them so that an anomaly is easy to spot. Take plenty of readings where there is a curve; for example, in an *I-V* graph for a light bulb, you should take a lot of readings where the filament is beginning to glow.

Use your instruments properly: keep your eye level with a thermometer; don't overtighten a micrometer; check for zero error; and most of all think about accuracy. Write your readings in your table to the correct number of significant figures (mark M1). Every measurement has a unit – so don't forget to include the units (mark M2).

Write down *all* your readings, and then work out the means for any repeated readings. Don't just write down a mean – the markers will want to see how you have worked out your mean.

Graphs

You need to plot a graph as part of your analysis. This will help you to spot a trend and then look at anomalies.

What to plot?

In physics it is best to try to plot data as a straight line, as this makes it easier to spot anomalies and deviations. This will usually require some manipulation of the data. To find out what graph to plot, you need to go back to the theory (marks A6 and A7). You can decide which graph you will plot in the planning section if you wish.

> **Worked Example**
>
> Find the **resistivity of a material**.
>
> ---
>
> If you are trying to find the resistivity of the material of a wire, then you will need to measure the resistance R of different lengths l of wire. The theory says that $R = \rho l/A$, where ρ is the resistivity and A is the cross-sectional area. So if you plot a graph of R against l, the line should be straight and pass through the origin, and the value of the gradient will be equal to ρ/A.

> **Worked Example**
>
> Measure the **acceleration of free fall**.
>
> ---
>
> You might use light gates or a mechanical timer to measure the time t an object takes to fall a distance h. Now from the theory, $h = \frac{1}{2}gt^2$, so you could plot a graph of h against t^2 (as shown on page 86 in the section on Gradients). If you compare $h = \frac{1}{2}gt^2$ with $y = mx + c$, you can see that h is equivalent to y and t^2 is equivalent to x, and the line should pass through the origin. The gradient m is equal to $\frac{1}{2}g$.
>
> Your final conclusion (mark A11) will include a value for g, which you get by finding the gradient and multiplying the value by 2.

Axes and labels

The dependent variable goes on the vertical axis. Labels should include a unit, so that a length l measured in centimetres is labelled l/m (mark A1).

Scales

Sensible scales for your graph (mark A2) should allow the data points to occupy as much of the paper as possible, at least half of the page up and down as well as across the page. It is sometimes better not to include the origin; you can always work out an intercept if you need to. For the scale, use multiples of 2, 4 or 5, otherwise you will find it very difficult to plot intermediate points.

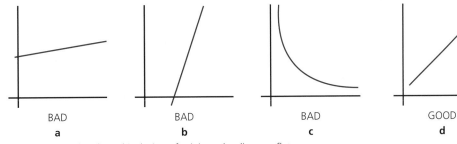

BAD	BAD	BAD	GOOD
a	b	c	d

a *Bad – vertical scale and inclusion of origin makes line too flat.*
b *Bad – horizontal scale and range makes line too upright.*
c *Bad – choose the variables to plot so you get a straight line.*
d *Good!*

Points

Your points need to be plotted accurately – to within half a square (mark A3). Use a sharp pencil and take a bit of time.

Line of best fit

Draw the line of best fit through the plotted points, either a straight line or a smooth curve (mark A4). There will almost certainly be a linear relationship between your variables, so your line of best fit will be straight. But you should not force it to go through the origin, as your data might show a systematic error. If the plots are above the line at each end and below in the middle, then you should be drawing a curve. Think before you draw! You should comment on any trend or pattern that you obtain (mark A5)

Gradients

You will probably need to find a gradient and/or an intercept from your graph to determine the relationship between your two variables (mark A6). Remember to use the line of best fit for working out the gradient, not the data points you have plotted. Continue your line of best fit to the edge of the page, and use as big a triangle as possible.

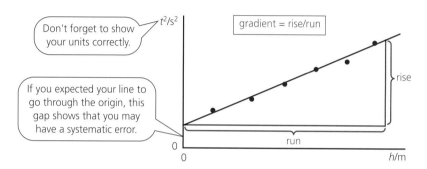

Quick Questions

Q3 How would you label a weight axis on a graph?

Q4 How does a graph help you reach a conclusion about your data?

Q5 If your line of best fit is not the straight line you were expecting, what might that tell you?

Evaluation

You need to evaluate your results to try to decide how reliable and accurate they are. Any candidate might comment on some of the aspects listed below, but better candidates say a little about them all.

Graph

Start by looking carefully at the line on your graph and considering the theory. If you were finding a value for something you already know (as in g by free fall), did you get an accurate value? Then look at the plots to see how scattered they are around the line. Is the line of best fit straight or does it deviate at the end? Your line might curve at the end due to the effect of air resistance, which only has a significant effect when the height dropped is quite big.

If your graph should go through the origin but doesn't, then you have a systematic error. You should try to identify a possible cause. In an experiment to find the velocity of sound, you might set up a standing wave in a vertical tube using a small loudspeaker. This method will always have an end correction and the graph will not go through the origin.

Uncertainties

This is the place to calculate your uncertainties (mark A10). It is usual to express uncertainties, errors and differences as percentages. In a school physics laboratory, you might expect to keep your uncertainties below about 10%. You should comment on the size of your uncertainties in relation to experimental errors.

Worked Example

Measure a **time**.

You measure the time T it takes for a trolley to run down a ramp. You record five readings: T/s = 4.53, 4.78, 4.29, 4.54 and 4.46. The mean is given by $T = 4.52\,s \pm 0.25\,s$ (this is half the range of the readings). The percentage uncertainty is given by $\delta T/T = 0.25/4.52 = 6\%$.

Worked Example

Measure a **current**.

When you are reading a current, the value fluctuates by 0.5 mA. One of your readings is 70 mA. This gives an uncertainty in the current of 0.5/70 = 0.7%. This is a very small uncertainty for school lab work.

Method

Now you should think about where errors might have crept in (mark A8). These might be in the theory – because you ignored air resistance, for example. Or they might be in the way you took your readings – it is hard to judge a resonance peak exactly and your reading might be just off the true value because of this.

Was your method a good one? In other words, were the readings easy to take? Did you get a good spread of readings, or could you improve on that? This is where you should suggest sensible improvements that would make your result more reliable or more accurate (mark A9).

Quick Questions

Q6 Explain how to calculate a percentage uncertainty in a value.
Q7 Show how you would calculate the percentage difference between two values.
Q8 Give an example of a systematic error.

ResultsPlus
Examiner tip

Don't write too much but do make sure you cover all the points in the marking grids.

Conclusion

Write down what you conclude from your practical work (mark A11), and mention the related physics principles (mark A7). You should finish by looking back at the aim of your experiment and comparing your outcome with what you were hoping to get. Try using common sense too and comparing your numerical value with something you know. You could also calculate the percentage difference. Suppose you got a value of $9.24 \, \text{m s}^{-2}$ for g. Then

ResultsPlus
Watch out!

Have you put units in all the right places?

$$\text{percentage difference} = \frac{(9.81 - 9.24)}{9.81} \times 100\%$$

That works out at just under 6% different, which is not bad for work in a school lab!

... and finally

You must make sure that this is all your own work, but since you will do all the work under supervision, this will not be difficult.

Don't forget

Now that you have read this make sure you take with you:

- calculator – has it got a new battery?
- ruler (30 cm clear plastic is best)
- sharp pencil (2B is about right), pencil sharpener and eraser
- pen or biro with black ink – take a spare too.

You may not take your own copy of the marking grids – or this book!

Answers

Answers to quick questions: Unit 1

Motion equations and graphs

1 The swimmer has swum a (scalar) *distance* of 150 m over 3 lengths, but at the end she is only 50 m away from her starting point so her (vector) *displacement* is 50 m along the pool.

2 a $v = u + at = 35\,\text{ms}^{-1} - 9.81\,\text{ms}^{-2} \times 5.0\,\text{s}$
$= -14.1\,\text{ms}^{-1}$ (i.e. it has passed its highest point and is on the way down again).
$s = ut + \frac{1}{2}at^2 = 35\,\text{ms}^{-1} \times 5.0\,\text{s} -$
$0.5 \times 9.81\,\text{ms}^{-2} \times (5.0\,\text{s})^2 = 52.4\,\text{m}$

 b Maximum height is achieved when $v = 0$.
$v = u + at$, so $t = (v - u)/a$
$= (0 - 35)\,\text{ms}^{-1}/ -9.81\,\text{ms}^{-2} = 3.57\,\text{s}$

Thinking Task

a

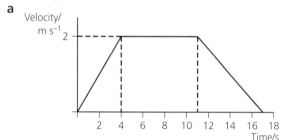

 b Area under graph $= (\frac{1}{2} \times 4\,\text{s} \times 2\,\text{ms}^{-1}) + (7\,\text{s} \times 2\,\text{ms}^{-1})$
$+ (\frac{1}{2} \times 6\,\text{s} \times 2\,\text{ms}^{-1}) = 24\,\text{m}$

 c $a = (v - u)/t = (0\,\text{ms}^{-1} - 2\,\text{ms}^{-1})/6\,\text{s} = -0.33\,\text{ms}^{-2}$

Combining and resolving vectors

1 a Your drawing should give you an answer between 11.5 and 11.8 ms^{-1} on a bearing 31° east of north

 b $v = \sqrt{(10^2 + 6^6)} = 11.66\,\text{ms}^{-1}$
$\theta = \tan^{-1}(6/10) = 31°$ east of north.

2 Horizontal component $= 8.0 \times \cos 38° = 6.3\,\text{ms}^{-1}$;
Vertical component $= 8.0 \times \sin 38° = 4.9\,\text{ms}^{-1}$

Thinking Task

a Your vector diagram should give you a final displacement of approximately 33 m at a bearing of 148°.

b They will finish in the same position.

Force and acceleration

1 Force from the acceleration $= m \times a = 70\,\text{kg} \times 1.5\,\text{ms}^{-2}$
$= 105\,\text{N}$

2 a $a = (v - u)/t = 15\,\text{ms}^{-1}/3.5\,\text{s} = 4.29\,\text{ms}^{-2}$
$F = m \times a = 109\,\text{kg} \times -4.29\,\text{ms}^{-2} = 467.6\,\text{N}$

 b The force calculated above is the force due to gravity. The cyclist must have been exerting this force to maintain a steady speed uphill, so when going downhill this force will be doubled. So
$a = F/m = (2 \times 467.6\,\text{N})/109\,\text{kg} = 8.58\,\text{ms}^{-2}$
$v = u + at = 0 + 8.58\,\text{ms}^{-2} \times 6.5\,\text{s} = 55.8\,\text{ms}^{-1}$

3 In freefall, the forces are initially unbalanced so the skydiver accelerates downwards until drag forces have increased enough to balance her weight. At this point she has reached terminal velocity. When she opens her parachute she increases the drag forces so she decelerates until the drag and weight balance again. Because the parachute has a larger area, the speed at which drag balances weight is much slower than when she was in free fall.

Thinking Task

Similarities	Differences
both forces are the same magnitude	forces in equilibrium act on the same object, but for a Newton's third law pair they act on different objects
the forces act in opposite directions	forces in equilibrium can be the same or different types, but Newton's third law pairs are always the same type of force

Gravity and free-body diagrams

1 D
$g_1 = F/m = 28\,\text{N}/8\,\text{kg} = 3.5\,\text{Nkg}^{-1}$
$g_2 = 2 \times 3.5 = 7.0\,\text{Nkg}^{-1}$
$F = m \times g = 16\,\text{kg} \times 7.0\,\text{Nkg}^{-1} = 112\,\text{N}$

2 a If the speed is constant the forces must be in equilibrium.

 b From your sketch you should see that the component of the weight acting along the slope $= \sin 35° \times 5\,\text{N} = 2.87\,\text{N}$
So friction $= 2.87\,\text{N}$ acting up the slope.

Thinking Task

3

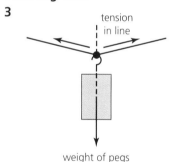

weight of pegs

The weight of the bag acts from its centre of gravity. The washing line is under tension, so the forces are equal in both directions. There must be an upward component of the force from the washing line to balance the weight of the pegs, so the line cannot be horizontal.

Projectile motion

1 Horizontal: time = distance/speed $= 0.8\,\text{m}/1.5\,\text{ms}^{-1} = 0.533\,\text{s}$
Vertical: $s = \frac{1}{2}at^2 = 0.5 \times 9.8\,\text{ms}^{-2} \times (0.533\,\text{s})^2 = 1.39\,\text{m}$

2 a Your sketch should be similar to that at the bottom of page 16.

 b $v^2 = u^2 + 2as$, so $u^2 = -2as = -2 \times -9.81\,\text{ms}^{-2} \times 2.5\,\text{m}$
$= 49.05$, so $u = 7\,\text{ms}^{-1}$

 c $\tan 65° = 7/u_h$, so $u_h = 7/\tan 65° = 3.26\,\text{ms}^{-1}$

3 Vertical component of velocity = $12\,\mathrm{m\,s^{-1}} \times \sin 33° = 6.54\,\mathrm{m\,s^{-1}}$

When the ball reaches the ground again, distance $s = 0\,\mathrm{m}$, and $v = -u$, $a = -9.81\,\mathrm{m\,s^{-2}}$

$v = u + at$, $-u = u - 9.81\,\mathrm{m\,s^{-2}} \times t$, $t = 2u/9.81\,\mathrm{m\,s^{-2}} = (2 \times 6.54\,\mathrm{m\,s^{-1}})/9.81\,\mathrm{m\,s^{-2}} = 1.33\,\mathrm{s}$

Horizontal distance = $12\,\mathrm{m\,s^{-1}} \times \cos 33° \times 1.33\,\mathrm{s} = 13.4\,\mathrm{m}$

Thinking Task

Air resistance will cause the object to decelerate faster than predicted, so it will not gain as much height and will reach its maximum height sooner. Air resistance will also reduce the downwards acceleration after this point, but the overall time of flight will still be less than predicted. The horizontal component of velocity will be reduced during the flight due to air resistance, so it will be moving more slowly (in a horizontal direction) during the part of the flight where it is also descending, so its flight path will not be symmetrical.

Work and power

1

with air resistance   without air resistance

a Horizontal component of force = $65\,\mathrm{N} \times \cos 50° = 41.8\,\mathrm{N}$
Work done = $41.8\,\mathrm{N} \times 85\,\mathrm{m} = 3553\,\mathrm{J}$

b Power = work/time = $3553\,\mathrm{J}/60\,\mathrm{s} = 59.2\,\mathrm{W}$

2 a $t = s/v = 12.5\,\mathrm{m}/0.3\,\mathrm{m\,s^{-1}} = 41.66\,\mathrm{s}$

b Power per passenger = $(700\,\mathrm{N} \times 12.5\,\mathrm{m})/41.66\,\mathrm{s} = 210\,\mathrm{W}$
For 50 passengers, power = $10\,500\,\mathrm{W}$ (or $10.5\,\mathrm{kW}$)

c Any three from:
The escalator is also moving its own weight, so would be doing more work.
There may be more/heavier passengers on it, or passengers with luggage.
The power of the motor would also have to overcome friction forces.
The motor is not 100% efficient, so it will require more input power than the power it transfers to the escalator.

Energy and energy conservation

1 a $\Delta E_p = mg\Delta h = 0.5\,mv^2$ so
$v = \sqrt{(2g\Delta h)} = \sqrt{(2 \times 9.81\,\mathrm{m\,s^{-2}} \times 0.015\,\mathrm{m})} = 0.54\,\mathrm{m\,s^{-1}}$

b Air resistance, and possibly friction in the suspension of the balls, will transfer some of the kinetic energy to heat and sound energy as the balls move.

2 a $E_k = 0.5 \times 1200\,\mathrm{kg} \times (25\,\mathrm{m\,s^{-1}})^2 = 375\,000\,\mathrm{J}$
Work done to stop the car = $F\Delta s$, so
$F = 375\,000\,\mathrm{J}/120\,\mathrm{m} = 3125\,\mathrm{N}$

b Find the acceleration: $v^2 = u^2 + 2as$, so
$0 = (25\,\mathrm{m\,s^{-1}})^2 + 2 \times a \times 120\,\mathrm{m}$
$a = -625\,\mathrm{m^2\,s^{-2}}/240\,\mathrm{m} = -2.6\,\mathrm{m\,s^{-2}}$
$F = ma = 1200\,\mathrm{kg} \times -2.6\,\mathrm{m\,s^{-2}} = -3120\,\mathrm{N}$
(differences are due to rounding intermediate values).

c The conservation of energy method involves fewer steps. The equations of motion produce an answer that also gives the direction of the force.

Thinking Task

Cricket pads or climbing helmets are intended to reduce the force of an impact on the body. The engineer needs to work out how well the equipment can convert the kinetic energy into other forms of energy (such as elastic strain energy), and so how much protection it will provide.

Fluids and fluid flow 1

1 a $3.4\,\mathrm{mm} = 0.0034\,\mathrm{m}$, volume = $3.93 \times 10^{-8}\,\mathrm{m^3}$

b Volume = $\frac{4}{3}\pi r^3 = 2.06 \times 10^{-2}\,\mathrm{m^3}$

c Radius = $0.056\,\mathrm{m}$, volume = $\pi r^2 l = 2.56 \times 10^{-3}\,\mathrm{m^3}$

2 a $m = 2700\,\mathrm{kg\,m^{-3}} \times 3.93 \times 10^{-8}\,\mathrm{m^3} = 1.06 \times 10^{-4}\,\mathrm{kg}$

b $m = 115.4\,\mathrm{kg}$

c $m = 1.54\,\mathrm{kg}$

3 a Upthrust = weight of fluid displaced = $5.2 \times 10^{-3}\,\mathrm{m^3} \times 1100\,\mathrm{kg\,m^{-3}} \times 9.81\,\mathrm{N\,kg^{-1}} = 56.1\,\mathrm{N}$

b Upwards – the upthrust is greater than its weight ($41.2\,\mathrm{N}$)

c The flow will be laminar, so your sketch should show smooth lines around the sphere.

Thinking Task

The more streamlined the shape, the higher the speed at which laminar flow becomes turbulent as air passes the car. Laminar flow produces less drag than turbulent flow.

Fluids and fluid flow 2

1 The higher the temperature, the less viscous the oil and so the lower the drag forces on an object falling in it. At higher temperatures the ball will have to be moving faster before the drag balances its weight, so its terminal velocity will be higher.

2 a Your sketch should show one upwards arrow for upthrust and two downwards arrows, one for weight and one for drag.

b Upthrust = weight + viscous drag, so
drag = $0.18\,\mathrm{N} - 0.17\,\mathrm{N} = 0.01\,\mathrm{N}$
$F = 6\pi r\eta v$ so $v = 0.01\,\mathrm{N}/(6\pi r\eta)$
$= 0.01\,\mathrm{N}/(6 \times \pi \times 1.8 \times 10^{-5}\,\mathrm{N\,s\,m^{-2}} \times 0.3\,\mathrm{m})$
$= 98.2\,\mathrm{m\,s^{-1}}$, upwards.

c This is a very high speed. The actual speed will be much less because there are other air resistance forces in addition to viscous drag.

Force and extension

1 A
$k = F/\Delta x = 5\,\mathrm{N}/4\,\mathrm{cm} = 1.25\,\mathrm{N\,cm^{-1}}$. Force on each spring = $10/3 = 3.33\,\mathrm{N}$.
$\Delta x = F/k = 3.33/1.25 = 2.67\,\mathrm{cm}$.

2 a Force to compress spring by $3\,\mathrm{cm} = (3/2) \times 9\,\mathrm{N} = 13.5\,\mathrm{N}$
E_{el} = area under force/extension graph, which will be a triangle base $3\,\mathrm{cm}$ and height $13.5\,\mathrm{N}$, so
$E_{el} = 0.5 \times 0.03\,\mathrm{m} \times 13.5\,\mathrm{N} = 0.203\,\mathrm{J}$

b $E_p = 0.203\,\mathrm{J} = mg\Delta h$, so
$h = 0.203\,\mathrm{J}/(0.018\,\mathrm{kg} \times 9.81\,\mathrm{N\,kg^{-1}}) = 1.15\,\mathrm{m}$

c The calculation ignores energy transfers due to air resistance.

Stress, strain and the Young modulus

1 a Cross section of wire = $\pi \times (0.42 \times 10^{-3}\,\text{m})^2$ = $5.54 \times 10^{-7}\,\text{m}^2$
Stress = $125\,\text{N}/5.54 \times 10^{-7}\,\text{m}^2 = 2.26 \times 10^8\,\text{N m}^{-2}$

b E = stress/strain, so strain = $2.26 \times 10^8\,\text{N m}^{-2}/3.4 \times 10^{11}\,\text{N m}^{-2} = 6.64 \times 10^{-4}$
Extension = strain × original length = $1.13 \times 10^{-4}\,\text{m}$

2 Equipment: wire, masses on hanger (100 g to 1 kg), clamp, wooden block, ruler, sticky tape to make marker on wire, micrometer, eye protection, box or bin.
Measurements: diameter of wire at several points and orientations measured, distance between block/clamp and marker, position of marker after each mass added.
Processing: find mean diameter of wire and then work out cross-sectional area, convert masses to forces by multiplying by 9.81, plot force against extension, draw line of best fit, find gradient. Find E from gradient × original length/cross-sectional area.
Precautions: wear eye protection in case wire snaps, place box or bin under the weights so they cannot fall on feet.

3 Both are a force per unit area.

Thinking Task

A wire does not stretch very much, so to make measurements with school lab equipment as accurate as possible the extension needs to be as large as possible. A thin wire is used to get a high stress for a given force, which will produce a larger extension than a thicker wire. For a given stress, there will be a bigger extension for a longer wire.

Answers to quick questions: Unit 2

Waves

1 a $2 \times 10^{-4}\,\text{s}$
b $0.066\,\text{m}$

2 Use $v = f\lambda$ to calculate the wavelength $\lambda = 0.8\,\text{m}$

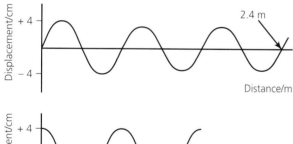

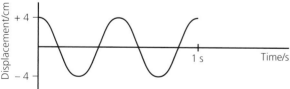

3 M will initially move down, then up, and be back at zero displacement after half a cycle.
L will move up to the maximum positive displacement.

Superposition and standing waves

1 Wavelength = 6 m and frequency = 55 Hz

2 Light waves from each slit meet at the screen. A path difference of 0, λ, 2λ, etc. results in a maximum (or blob of light) as the waves meet in phase. A path difference of $\lambda/2$, $3\lambda/2$, etc. results in a minimum as the waves are in antiphase.

3 There will be an antinode at each end of the pipe. The first (fundamental) standing wave pattern will have a node in the centre of the pipe. The first overtone will have three antinodes and two nodes.

Thinking Task

a The ray travels to the bottom of the oil layer and back.
b A maximum because the waves arrive in phase
c The oil layer will have different thicknesses. The path difference will correspond to different wavelengths, i.e. different colours of light.

The electromagnetic spectrum

1 Wavelength = 1515 m. Waves diffract best past objects or through gaps of size comparable with wavelength. These radio waves will diffract (spread) well past hills into valleys, etc.

2 The UV light is not visible to the naked eye.

3 Your answer could include things such as: X-rays of bodies or airline luggage; IR to detect bodies in rubble/criminals at night; or IR, UV, etc. to detect objects in space.

Thinking Task

If the waves meet, then superposition occurs. If the phase difference between the waves remains constant (coherent), then interference effects will remain constant and therefore noticeable.

Pulse–echo techniques

1 a A reflected wave pulse is timed. The distance it travels (distance to the surface) is calculated using distance = speed × time.
b Shorter pulses and higher frequency/shorter wavelength

2 The Doppler effect means that reflected waves have a different frequency (or wavelength) from the emitted wave. The larger the difference, the faster the car.

3 a Distance travelled to far side and back = $1400\,\text{m s}^{-1} \times 6.6 \times 10^{-5}\,\text{s} = 0.0924\,\text{m}$
Thickness of organ 0.046 m
b The gel reduces the reflection as the sound wave meets the skin as it excludes air.

Refraction

1 The completed ray is a line vertically downwards.
2 a $2.1 \times 10^8\,\text{m s}^{-1}$
b 27°
c

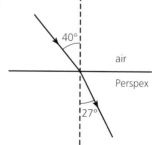

Thinking Task

a A normal needs drawing from centre of drop through point where ray meets the drop. Angle i is about 52° and angle r is about 36°.
b Critical angle = 50°
c Draw a normal to the point at the back of the drop. The angle between the ray and the normal at this

point is 40°. This is less than the critical angle so it is partially reflected.

d The refractive index for different colours has different values.

Polarisation

1 The light from the sky is polarised. When the oscillations are in line and absorbed by the filter the sky appears dark.

2 The light from the road surface is polarised. If the polarising filters in the glasses are at the correct angle (in the plane at right angles to the plane of polarisation of the reflected light), they will absorb this reflected light.

3 D

Thinking Task

a Polarised – oscillations are in one plane only. Unpolarised – oscillations are in many planes.

b Segments become polarising filters and will absorb one plane of oscillations of light. If the light is polarised in that plane, then these segments appears dark and an image is formed.

c The whole screen appears dark.

Diffraction

1 The X rays have a wavelength approximately equal to the separation of atom layers in the quartz crystal.

2 Atoms in the unstretched band are not layered (ordered).
The atoms in the stretched band are ordered.

3 a The slits have to be narrow so that the light diffracts (spreads) as it passes through each slit, and then the two light waves overlap and interfere.

b If the slits are too far apart, the rays will not overlap.

Thinking Task

Electrons can be diffracted and produce interference patterns. This can be explained by considering waves as electrons. Neither student is wrong. Electron can exhibit wave and particle behaviour.

Current and potential difference

1 $0.24\,C$

2 $540\,J$

Thinking Task

The emf is the energy supplied to each coulomb of charge.
Potential difference is the energy transferred by each coulomb of charge between two points in a circuit.

Current-potential difference graphs

1 $24\,\Omega$

2 $2\,\Omega$

3 $24\,\Omega$

Thinking Task

As the filament heats up, the atoms vibrate more vigorously and there are more collisions between the electrons and atoms, increasing the resistance.

Power and work in electric circuits

1 a $8\,A$
  b $28.8\,\Omega$

2 $12\,V$

3 $5\,minutes$

Resistance and resistivity

1 a Total resistance = $0.5\,k\Omega$
    Current in the $2\,k\Omega$ resistors = $5\,mA$
    Current in the $1\,k\Omega$ resistor = $2.5\,mA$

  b Total resistance = $2\,\Omega$
    Current in the $2\,\Omega$ and $1\,\Omega$ resistors = $2\,A$
    Current in the $6\,\Omega$ resistor = $1\,A$

2 Drift velocity = $1 \times 10^{-5}\,m\,s^{-1}$

3 $10^{29} \times 10^{-8} = 10^{21}$
  $10^{18} \times 10^3 = 10^{21}$
  This suggests that number density is inversely proportional to resistivity.

Thinking Task

Measure length of sheet = $30\,cm$ approx

Width = $21\,cm$ approx

Thickness of 50 pages is $3\,mm$ approx so thickness of one sheet is $0.06\,mm$

Cross-sectional area = $1.3 \times 10^{-5}\,m^2$ approx

Resistance = $0.8\,\Omega$ approx

Potential dividers and internal resistance

1 a $0.5\,A$ b $0.15\,V$ c $1.35\,V$

2 a $0.25\,A$ b $1.0\,V$ c $1.5\,V$

  d $0.5\,V$

Thinking Task

There is a sudden demand for a much larger current. The 'lost volts' across the battery is $0.040\,\Omega \times 100\,A = 4\,V$. The terminal e.m.f. drops to $8\,V$, so the headlights dim.

Photoelectric effect

1 $5.70 \times 10^{14}\,Hz$

2 If the frequency is less than the threshold frequency, the photon energy is less than the work function, so no photoelectrons are emitted. The sodium will absorb the energy of the photons and heat up slightly.

3 $9.01 \times 10^{-20}\,J$

Thinking Task

Graph starts further from the origin but still on the negative V axis (takes less energy for the photoelectrons to be released, so they have more KE, which leads to a greater stopping potential).

Photons, spectra and energy levels

1 $4.86 \times 10^{-7}\,m$

2 An emission spectrum is produced when an electron in an atom is excited to a higher energy level and then drops back to a lower energy level, giving out energy in the form of a photon. This is seen as a line at a particular frequency in the spectrum from an element. An absorption spectrum is due to the absorption of a photon by an atom, resulting in an electron in the atom moving from a lower energy level to a higher energy level. This produces a dark line in a continuous spectrum, at the frequency corresponding to the photon absorbed.

3 The atoms in each chemical element have a unique set of energy levels. When photons are emitted or absorbed at frequencies that correspond to the differences between energy levels, a line spectrum unique to that element is produced.

Thinking Task
Wavelength is inversely proportional to frequency, so photon energy decreases as wavelength increases.

Radiation flux and solar cells
1   16%
2   3.2 W
3   a   20 J each second
    b 5.0×10^{19} photons per second

Thinking Task
a   Area = $(4000\,\text{W}/1000\,\text{W m}^{-2}) \times 4 = 16\,\text{m}^2$
b   Assume that radiation flux is the same as at the surface of the Earth (it may be more than this, as there is no attenuation from the atmosphere), and that the cells are 25% efficient (which modern cells can achieve).

Answers to quick questions: Unit 3

Thinking Task
Based on the marking grids, headings could be: Venue, What we saw, The physics, Why it is there, The practical work.
1   a   The variable you change.
    b   The variable you measure as a result, the outcome.
2   If you are measuring resistance and length the control variables are the material the wire is made from, temperature and the cross-sectional area of the wire.
3   W/N
4   It shows a trend in your data, this is probably a linear relationship between the variables (straight line). If it passes through the origin it is a proportional relationship. A graph can be used to calculate values of quantities not measureable directly, like Young modulus. It shows up anomalous readings. If the line is only straight over a certain range it shows the regions where the relationship is not true.
5   The theory is invalid or needs correcting (for air resistance perhaps).
6   Take the range (or half the range) in the readings, as the actual uncertainty, multiply this by 100 and divide by the mean value.
7   The percentage difference is the difference between the two values multiplied by 100 and divided by the mean of the two values.
8   A set of scales that reads 0–10 N when there is a non-zero value nothing on them. An instrument that reads a non-zero value when it is measuring nothing, such as a balance that reads 0.1 N when nothing is being weighed.

Answers to practice exam questions: Unit 1

Section 1
1   B   2   C   3   D   4   B
5   a   i   You should have calculated the area between the line on the graph and the time axis, for the time from 0 to 1 second [1]
       ii   Distance travelled from 1 to 4 seconds = area between line and time axis = 45 m [1]
          Total distance = 50 m [1]

b   Rose 5 m, fell 5 m [1] displacement = 40 m below point of release [1]
c   Your line should be parallel to the time axis, from t = 0 s to t = 4 s [1], and it should be labelled as the acceleration being **minus** [1] $9.81\,\text{m s}^{-2}$ [1]
6   a   i   Use $s = \tfrac{1}{2}at^2$ to calculate that acceleration = $1.73\,\text{m s}^{-2}$ [1]
       ii   No air resistance [1], so resultant force on objects does not change [1]
    b   Use $W = mg$ [1] to calculate W = 179 N [1]
    c   i   Calculate vertical component of velocity = $15.4\,\text{m s}^{-1}$ using 45 sin 20° [1] Use $v = u + at$ [1] to calculate t = 18.1 s [1]
       ii   Calculate horizontal component of velocity = $42.3\,\text{m s}^{-1}$ [1] using 45 cos 20° Distance = 766 m [1]
       iii   This is only about half a mile [1]
7   a W = force × distance [1] = $110\,\text{kg} \times 9.81\,\text{N kg}^{-1} \times 2.22\,\text{m}$ = 2395.6 J [1]
    b   Power = work done/time [1] = 2396 J/3 s = 798.6 W [1]
    c   Energy can neither be created [1] nor destroyed [1] (just saying 'energy is conserved' gets no marks)
    d   i   Chemical energy (in the body of the weightlifter) or work done (lifting bar) = (gain in) GPE (of bar) [1]
       ii   Transfer from GPE to KE [1] GPE lost = KE gained [1]
    e   Setting $\tfrac{1}{2}mv^2 = mgh$ or $\tfrac{1}{2}mv^2$ = work done or 2400 J [1], correct values substituted [1], answer: $6.6\,\text{m s}^{-1}$ [1]

Section 2
1   D
2   A
3   D
4   a   High-viscosity flow would be slower than low-viscosity flow [1]
    b   The distance travelled by the lava in a set time (or the speed of each lava flow) [1]
    c   Cooling increases the viscosity [1]
    d   A diagram of laminar flow should have at least 2 smooth lines [1] that do not cross [1]. Turbulent flow should show whirls or eddies [1] in at least 3 lines.
    e   Use a log scale/powers of 10 scale [1]
5   a   Use $F = \sigma \times A$ [1] to work out $F = 1.6 \times 10^{10}$ N [1]
    b   Use $\varepsilon = \sigma / E$ [1] to work out strain = 0.16 [1]
    c   Brittle [1]
6   a   Elastic [1], as it must return to its original length when the force is removed [1]
    b   Points plotted correctly [1] to within plus or minus half a square [1], and a straight line of best fit [1]
    c   Use any two pairs of points from the table or graph through origin [1] to work out stiffness = $0.53\,\text{N mm}^{-1}$ [1] (or $530\,\text{N m}^{-1}$).
    d   3.2 N [1]
    e $E_{el} = \tfrac{1}{2}F\Delta x$ [1] = 9.6×10^{-3} J [1].
    f   Half the original force [1]

Answers to practice exam questions: Unit 2

Section 3

1 C   **2** A   **3** B

4 a i 85 [1]
   ii First calculate BP from Pythagoras; BP = 278 cm so 87 wavelengths [1]

b Maxima at P [1] as waves in phase/path difference whole number of wavelengths to give constructive interference [1] intensity then decreases as waves get progressively out of phase [1] till minima when waves exactly out of phase to give destructive interference [1]

5 a Travelling wave (OR travelling vibration) (on string) [1] Wave reflects at the end (OR bounces back) [1] Incident and reflected waves (OR waves travelling in opposite directions) superpose (OR interfere OR combine) [1]

b Use of node to node distance = $\lambda/2$ / recognise diagram shows 2λ [1]
Correct answer 0.8 m [1]

c Use of $v = f\lambda$, [1] to give 18 m s [1]

d Node [1]

e i Large amplitude motion [1] (OR same speed/ frequency)
   ii Antiphase/180° or π radians [1]

f Use of $v = f\lambda$ to give 0.6 m [1]; diagram showing two whole wavelengths on the string [1]

6 a Transverse waves oscillate in any direction perpendicular to wave direction [1]. Longitudinal waves oscillate in one direction only (OR parallel to wave direction) [1]. Polarisation reduces wave intensity by limiting oscillations and wave direction to only one place (OR limiting oscillations to one direction only) [1]. (Accept vibrations and answers in terms of an example such as a rope passing through slits.)

b Light source, 2 pieces of polaroid and detector e.g. eye, screen, LED OR laser, 1 polaroid and detector [1]
Rotate one polaroid [1]
Intensity of light varies [1]

Section 4

1 C

2 C

3 a i $R = V/I$ [1] = 2.68 V/0.31 A = 8.6 Ω [1]
   ii Lost volts across battery 2.80 V – 2.68 V = 0.12 V [1]
Internal resistance = 0.12 V/0.31 V = 0.39 Ω [1]

b i Charge = It [1] = 2 A × 3600 s = 7200 C [1]
For two cells the total charge is 14 400 C [1]
   ii $t = Q/I$ [1] = 14 400 C/0.31 A = 4.7 × 10^4 s [1]

c Efficiency is power output/total power input = I^2R/ I^2 $(R+r)$ [1]
where R is the resistance of the torch bulb and r is the internal resistance of the battery.
As r increases, then the efficiency is less [1]

Section 5

1 B

2 a Arrow showing electron moving from lower level to a higher level [1]. Arrow downwards from higher to lower level (must show smaller energy change than upward arrow) [1]

b Minimum energy when $\lambda = 400 \times 10^{-9}$ m [1]
Use of $f = c/\lambda$ [1]
Use of $E = hf$ [1]
Correct answer (3.1 eV) [1]

3 a Waves spread out when passing through a gap / past an obstacle [1], λ stays constant [1]

b Diagram showing 2 waves in phase [1] adding to give larger amplitude [1]

c Atomic spacing (similar to λ)
Regular / ordered structure
Symmetrical structure
DNA is a double helix structure [2] (Max 2)

d Behave as waves [1]

Answers to Unit 1: Practice unit test

1 A   **2** A   **3** B

4 C   **5** B   **6** D

7 a A increases, B stays the same [1]
b A and B both stay the same [1]
c A decreases, B stays the same [1]
d A and B both stay the same [1]

8 a Use $E_p = mgh$ [1] to work out E_p = 37 278 J [1]

b i E_p gained by projectile = 10 800 J [1]
E_k = 37 278 J – 10 800 J = 26 478 J [1]
(or 26 200 J if 37 000 J is used)
   ii All the lost gravitational potential energy has been transferred to kinetic energy of the projectile and counterweight (or the mass of the moving arms is negligible) [1]
   iii Any two points from: E_k of counterweight; $E_k = \frac{1}{2}mv^2$; counterweight has speed $v/4$; due to the lever arm ratio 1 : 4 [2]

c i Use $s = ut + \frac{1}{2}at^2$ [1] to work out t = 2.07 s [1]
   ii Use $s = vt$ [1] to work out s = 46.6 m [1]

9 a Reaction time of the driver and cyclist [1]
b i 6.8 or 6.9 seconds [1]
   ii Area under the line for the car up to 6.8 s = 27 m [1]
Area under the line for the cyclist up to 6.8 s = 45 m [1]
Distance = 18 m [1]
c They are the same [1]

10 a Use $s = ut + \frac{1}{2}at^2$ or $mgh = = \frac{1}{2}mv^2$ and other equation(s) [1] t = 0.69 s to 2 sig fig [1] (allow g = 10 m s^{-2} throughout)
b Select 2 correct equations [1] substitute correct values [1] t = 0.38 s [3]
c Use $v = d/t$ [1] to find v = 18.9 m s^{-1} [1]
d Any from friction/drag/inelastic collision at bounce [1]

11 a i Laminar [1]
   ii At least two streamlines in front of the skier [1], at least two streamlines going smoothly around the skier (they must not cross or touch) [1]
   iii Any from: smooth/tight-fitting/elastic [1]
b Desirable: elastic (will return to original shape when force removed) or tough (can withstand impacts without breaking) [1 for property, 1 for explanation for property]
Undesirable: plastic (will remain deformed once load removed) [1 for property, 1 for explanation]

12 a Strong – large force/load/stress required to break [1]
Brittle – shatters/cracks/breaks with little plastic deformation [1]
Plastic deformation – does not return to original length when load removed [1]

b Use $A = \pi r^2$ [1] and $F = \sigma A$ [1] to work out force = 193 N [1]

c Use $\varepsilon = \sigma/E$ [1] to work out strain of 0.015 [1], then extension 0.017 m [1]

13 a i Points plotted correctly [1] both axes to be labelled with quantities and units, and drawn the correct way round [1]

ii Line of best fit drawn (must be thin, continuous) [1]
Use $k = F/x$ [1] to calculate $k = 22\,N\,m^{-1}$ [1]

b $E = \frac{1}{2}Fx$ [1] to give $E = 514\,J$ [1]

c Any of: rope is connected to sheet as many parallel sections, making it stiffer/each section of rope is supporting much less than the full weight/there may be energy lost to friction between the rope and the trampoline frame [1]

d It would have to withstand greater stress [1], so the rope would need to be thicker or shorter to reduce the extension [1]

Answers to Unit 2: Practice unit test

1	D		2	A	3	C		4	C
5	B		6	D	7	B			

8 a Use of $\mu = \sin i/\sin r$ [1], 20° [1] (allow 20–21°)
b Use of $\mu = 1/\sin C$ [1] 50° (allow 50–51°)
c Angle of incidence = 35° (allow 33–37°) [1]
Ray of light shown refracting away from normal on leaving raindrop [1]
Some internal reflection shown with $i = r$ [1]
Reflected ray shown refracting away from normal as it leaves the front of the raindrop / angle of refraction correctly calculated at back surface [1]

9 a Oscillations in one plane only [1]
b Use a polarising filter and rotate to cut out the polarised light [1]. Insert the sugar solution and adjust the polarising filter [1]. Measure the angle through which it had to rotate with protractor [1].

10 a The atoms are ordered [1] and the gaps between layers are approximately equal to the wavelength of electron beam [1]
b The electrons behave like waves [1]

11 Any 5 of: Lamps X and Y are equal brightness with switch open
Lamp X and Z will be equal in brightness
Lamp Y will become brighter than before
Lamp X and Z will be less bright than X was before
The current splits through X and Z
The potential difference is less across X and Z than Y.
The combined resistance of X and Z is less than (half) resistance of Y
Power dissipated = current × potential difference

12 Any 4 of: There are volts lost across the internal resistance. The rest of the voltage is available to the rest of the circuit (headlamps). The larger the current drawn from a battery, the larger the volts lost. When the starter motor is on, there is a large current. The lost volts increases / the potential difference across headlights is smaller.
Once engine is running, it generates 12 V for headlights.

13 Max 5, from:
Reference to $I = nqvA$ [1]
For the lamp
Increased atomic vibrations reduce the movement of electrons [1]
Resistance of lamp increases with temperature [1]
For the thermistor
Increased atomic vibrations again reduce movement of electrons [1]
But increase in temperature leads to a large increase in n [1]
Overall the resistance of the thermistor decreases with increase in temperature [1]

14 a Electrons within atoms excited to higher energy levels [1] move down from one energy level to another, emit photons/energy/em radiation [1]
b i Each frequency/wavelength/photon energy is seen as a separate/discrete line (of a different colour) [1]
ii Recall of $v = f\lambda$ [1] $5.1 \times 10^{14}\,Hz$ [1]
c Neon atoms have different energy levels for electrons [1]; the difference in these energy levels corresponds to different energies/wavelengths of photons [1]; different colours of light correspond to different energies/wavelengths [1].
d The oscillations are at right angles to the direction of the wave [1]

15 a Waves are (partially) reflected if they encounter a change of medium [1]
b If the densities are very different, the reflected intensity is greater [1]
As the wave penetrates deeper, then there will be less energy so less intensity [1]
c The distance to the next interface between media is different [1]
d i The frequency of the returning wave is different from the original wave [1], because a moving reflector either bunches waves together or stretches them apart [1]
ii If the reflector is moving away, then the wavelength stretches or increases [1]. The frequency decreases [1] because $f = v/\lambda$ [1]

16 a i Constant phase difference [1]
ii Energy is not transferred [1], produced by waves travelling in opposite directions [1]
b Antinode in the middle of a hot zone [1]
c Use of $v = f\lambda$ [1] 12.3 cm [1]
d Path difference = 6.1 cm [1]
e The wavelength is 12 cm [1], so the path difference is half a wavelength [1]. You would expect a minimum or node [1]
f The antinodes and nodes would be in different positions for the different frequencies [1]
Therefore a cold spot for one frequency might be a hot spot for the other [1]

17 a Ohmmeter **only** connected to both contacts or ammeter in series [1], voltmeter in parallel and battery [1]
b Use of $R = \rho l/A$ [1], substitute correct values including $A = \pi r^2$ [1], 172 Ω [1]

Index